Roblox Game Development

The **Official**
R⬛BL⬛X Guide

in **24** **Hours**

Pearson

Roblox Game Development in 24 Hours: The Official Guide

ISBN-13: 978-0-13-682973-7
ISBN-10: 0-13-682973-2

Library of Congress Control Number: 2021931227

1 2021

Trademarks

All terms mentioned in this book that are known to be trademarks or service marks have been appropriately capitalized. Pearson cannot attest to the accuracy of this information. Use of a term in this book should not be regarded as affecting the validity of any trademark or service mark.

Warning and Disclaimer

Every effort has been made to make this book as complete and as accurate as possible, but no warranty or fitness is implied. The information provided is on an "as is" basis. The author and the publisher shall have neither liability nor responsibility to any person or entity with respect to any loss or damages arising from the information contained in this book.

Special Sales

For information about buying this title in bulk quantities, or for special sales opportunities (which may include electronic versions; custom cover designs; and content particular to your business, training goals, marketing focus, or branding interests), please contact our corporate sales department at corpsales@pearsoned.com or (800) 382-3419.

For government sales inquiries, please contact governmentsales@pearsoned.com.

For questions about sales outside the United States, please contact intlcs@pearson.com.

Executive Editor
Debra Williams Cauley

Acquisitions Editor
Kim Spenceley

Editorial Services
The Wordsmithery LLC

Managing Editor
Sandra Schroeder

Senior Project Editor
Tonya Simpson

Copy Editor
Charlotte Kughen

Indexer
Cheryl Lenser

Proofreader
Sarah Kearns

Editorial Assistant
Cindy Teeters

Cover Designer
Chuti Prasertsith

Compositor
Bronkella Publishing LLC

Contents at a Glance

Table of Contents

Foreword

Imagine a virtual universe built by a global community of artists, coders, storytellers, and everything in between. In this dream, people from all corners of the world come together to create and share millions of experiences with their friends and learn from one another. It would be a universe driven by imagination, where anything could be made and experienced, regardless of device, location, or time period. What if I told you this digital utopia has been a reality for over a decade?

When Erik Cassel and I co-founded Roblox in 2004, our vision was to create an immersive, 3D, multiplayer, physically simulated space where anybody could connect and have fun doing things together. In the early days of Roblox, we were fascinated by what people were making. We saw experiences where people wanted to manage their own restaurant, survive a natural disaster, or imagine what it's like to be a bird. Seventeen years later, as I gaze into the future, it's obvious this platform can become so much more.

Roblox is ushering in a new category of human co-experience, blurring the lines between gaming, social networking, toys, and media. Our team has found that the millions of daily Roblox users aren't just logging on to play games but are coming together to build communities, stories, and experiences with friends and strangers alike.

As we continue our mission to build a human co-experience platform that enables shared experiences among billions of users, there has never been a better time to join a global community of creative individuals who are contributing such amazing works to our platform. Developing 3D experiences is not only fun, but it also provides the skills and knowledge to launch a career in computer science, design, art, and so much more. Many top developers on our platform have used the money they earned from their creations on Roblox to pay for their college tuition, start their own game development studios, or put a down payment on a house for their parents.

I believe that ultimately Roblox will lead us to the creation of the Metaverse, a full-fledged digital reality that will complement our physical one. We can start to imagine a day where people aren't just coming to Roblox to play and socialize but also to hold business meetings or go to school. As the possibilities of the Metaverse increase by the day, so too does the need for innovative and creative developers who can shape the experiences we've been dreaming about in science fiction for years.

I personally invite you to join the world of Roblox not just as a player but also as a creator. Learning to develop both games and immersive 3D experiences can help connect millions of people worldwide through the power of play and create a community not defined by borders, languages, or geography. If you're at all interested in coding, game design, or the immersive 3D world of Roblox, consider peering through these pages and embracing your wildest, most creative ideas. The Metaverse depends on creators just like you.

Your imagination awaits,
David "Builderman" Baszucki
Founder + Chief Executive Officer
Roblox Corporation

About the Author

 **Genevieve Johnson** is the senior instructional designer for Roblox, the world's largest user-generated social platform for play. In her role, she oversees creation of educational content and advises educators worldwide on how to use Roblox in STEAM-based learning programs. Her work empowers students to pursue careers as entrepreneurs, engineers, and designers. Before working at Roblox, Johnson was educational content manager for iD Tech, a nationwide tech education program that reaches more than 50,000 students ages 6 to 18 each year. While at iD Tech, she helped launch a successful all-girls STEAM program, and her team developed educational content for more than 60 technology-related courses with instruction on a variety of subjects, from coding to robotics to game design.

About the Contributors

Ashan Sarwar is a Roblox developer who has been using Roblox Studio since 2013. He is the owner of LastShot, a Roblox shooting game on Roblox.

Raymond Zeng is a Roblox developer who loves programming and teaching all levels of programmers. He has a YouTube channel under the name of MacAndSwiss where he teaches Lua, talks about Roblox news, and showcases his programming projects.

Theo Docking has been working as a gameplay programmer for four years. He likes working on exciting projects, pushing Roblox to the limit, and meeting amazing people along the way. He loves playing with Roblox's physics engine and writing back-end code for NPCs, cars, and more. When he's not writing code, he's drawing up game design plans or playing Ultimate Driving to get some fresh air.

Joshua Wood discovered Roblox in 2013 and started making his own games a year later. He is the developer of Game Dev Life, which has had more than a million play sessions. He's also the owner of DoubleJGames.

Swathi Sutrave is a self-professed tech geek. She has been a subject matter expert for several different programming languages, including Lua, for corporations, start-ups, and universities.

Henry Chang is a computer graphics artist who practices in multiple mediums, including 3D, 2D, graphics, and animation. He is a self-starter and has been involved in interactive media opportunities. For more information, visit https://www.henrytcgweb.com/.

We Want to Hear from You!

As the reader of this book, *you* are our most important critic and commentator. We value your opinion and want to know what we're doing right, what we could do better, what areas you'd like to see us publish in, and any other words of wisdom you're willing to pass our way.

You can email or write to let us know what you did or didn't like about this book—as well as what we can do to make our books better.

Please note that we cannot help you with technical problems related to the topic of this book.

When you email, please be sure to include this book's title and author, as well as your name, email address, and phone number. We will carefully review your comments and share them with the author and editors who worked on the book.

Email: community@informit.com

Reader Services

Register your copy of *Roblox Game Development in 24 Hours: The Official Guide* at www.informit.com/register for convenient access to downloads, updates, and corrections as they become available. To start the registration process, go to informit.com/register and log in or create an account.* Enter the product ISBN (9780136829737) and click Submit.

*Be sure to check the box that you would like to hear from us to receive exclusive discounts on future editions of this product.

HOUR 1
What Makes Roblox Special?

What You'll Learn in This Hour:
▶ How Roblox empowers social connectivity
▶ How Roblox manages user content
▶ How Roblox enables fast prototyping and iteration
▶ What's inside Roblox's engine

Welcome to Roblox! With this free online 3D platform and game creation system, your only limitation is your imagination. Anyone can create and play unique games in a vibrant and interconnected social environment. There are millions of worlds to choose from, all created by users who bring their own unique style of gameplay, design, and communities. In this hour, you'll learn about the features Roblox provides for you to begin making your own expertly crafted game. Knowing the capabilities of Roblox will set you in the right direction.

Roblox is an all-in-one platform. In other engines, being a developer means you need engine code, moderation, and more. Luckily, Roblox handles all of that for you. Its infrastructure takes care of the dirty work, including server hosting and multiplayer networking, which leaves you with more time to create. A single account is all you need both to play and create.

The Roblox platform provides a built-in audience of millions of daily visitors with built-in moderation, cross-platform play, and a hard-coded currency called Robux. This means developers can focus on what matters most, creating new engaging experiences for players around the globe (Figure 1.1).

FIGURE 1.1
Welcome to Roblox.

To publish games, you don't need years of coding experience or expensive software licenses, and you don't have to complete a lengthy submission process. All you need is a PC or Mac, a stable Internet connection, Roblox Studio, and a bit of imagination, and your game can be published in minutes—for free.

As your game and player base grows, you can exchange the Robux you earn through player purchases in your game for real-life cash.

Roblox Empowers Social Connectivity

On Roblox, social connectivity is valued. Games can range from highly detailed hangout places to competitive edge-of-your-seat gaming experiences.

Also, because Roblox is cross-platform, anyone can play, whether they are joining on a PC, Mac, or even a mobile device. No one is excluded. To encourage social connection, Roblox provides the ability to add friends, chat together, and emote. Together, players explore worlds, solve puzzles, and even watch performances by their favorite musicians.

Roblox as a Social Website

Users in Roblox can create **Groups** to organize their social experience. Each Group has a page (Figure 1.2) where users can host their own games, sell virtual merchandise, and associate with other Groups, all while maintaining a unique identity. It's also common for Groups to be used as a brand hub where teams can combine their resources for development.

FIGURE 1.2
A game development Group page with their external social media links.

Roblox as a Creator Hub

Roblox provides ways for developers to connect with each other. The Roblox Developer Forum is an online forum where Roblox developers can discuss topics and exchange services. To join the Developer Forum, you just need to regularly browse and read the content in it. After you've spent enough time browsing and reading the resources, you will automatically level up (from visitor to member and even to community sage) and can start making your own posts.

Roblox has a real-life invite-only convention: the Roblox Developers Conference. At this conference, featured developers and members of the official Roblox team give speeches and presentations about their advances on the platform. You're not alone on your development journey!

The *Annual Bloxy Awards*, a yearly awards livestream and game-based ceremony, is another event that boosts the identity of Roblox as a social game. Roblox creators are nominated and awarded based on the community's vote in this event to win a virtual one-of-a-kind trophy. You can read more about these events in the official Roblox Blog (Figure 1.3) at https://blog.roblox.com/ and the DevForum.

FIGURE 1.3
A blog post featuring the 6th Annual Bloxy Awards in 2019.

Roblox Manages User Content

Roblox is very hands-free on the concepts that players can design, so you have the freedom to let your imagination run wild. Just about everything on Roblox is tied to a user account; games, cosmetics, plugins, and game assets are only some of the things that users can upload. Users have nearly complete freedom to decide what they want in their games.

NOTE

All Uploaded Assets on Roblox Must Pass Moderation

All content on Roblox must pass through a moderation check before it's allowed to be displayed to players. Users can also report anything deemed inappropriate so it can be flagged and deleted. This is not limited to your games and assets, but your account as well. It is encouraged that you read Roblox's rules and Terms of Service (https://en.help.roblox.com/hc/en-us/articles/115004647846-Roblox-Terms-of-Use) in the support section to learn more about moderation.

Organizing Content

Roblox has built-in organization for certain assets and products uploaded onto the website. Things you made will be in your Create page (Figure 1.4) and everything you obtained will be in your Inventory.

FIGURE 1.4
The Create page, including Developer Resources with links for Roblox Studio, the official Developer Wiki (Docs), and the DevForum (Community).

From the Create page, you can also access the Asset Library (Figure 1.5), which provides things such as Models, Decals, Audio, Meshes, and Plugins.

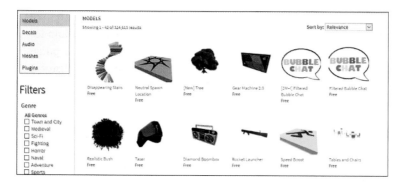

FIGURE 1.5
The Asset Library, where assets are on display.

Creating Your Identity

Don't forget that Roblox is a social website. Developers can create and upload images such as icons, thumbnails, and even advertisements to display on the site (Figure 1.6). You can use external tools of your choice to create a custom visual and upload it. These features are what gives groups, games, and characters on Roblox a diverse set of identities.

FIGURE 1.6
User-uploaded images on *Arsenal*'s game store tab for micro-transactions (*Arsenal* by ROLVe Community).

Customizing Your Characters

The Avatar Shop, also called the Catalog (Figure 1.7), is where users can purchase virtual items for their avatars, such as hats, heads, gear, accessories, and other items. Although the official Roblox account has been responsible for Avatar Shop content, the creation of shirts, t-shirts, and pants have always been in community hands. They have been considered core parts of fashion design groups and identity-focused clubs. Because Avatar Shop content items are assets with their own IDs, developers can load them onto Studio for their games.

FIGURE 1.7
Categories and some cosmetics on the catalog of the Avatar Shop.

Figure 1.8 shows a player's inventory featuring sections of catalogs for what they made and own, such as assets from the Asset Library and Avatar Shop.

FIGURE 1.8
A player's inventory.

Starting in August 2019, select users have been able to upload their own hats in the Accessories category, making it the first instance of users being able to upload hats. (Over time, this feature will be rolled out to more users.) Some hats require Premium to purchase.

Roblox has two official avatar rigs: the legacy R6 and the newer R15 Rthro. Both may use Avatar Shop cosmetics. Custom rigs may also be uploaded via Models but may require editing in-engine with Roblox Studio before being fully compatible for your game.

Roblox Enables Fast Prototyping and Iteration

Roblox is a flexible engine that maximizes your time to create. Roblox games use a language called Lua that requires no compiling. You can switch from coding to testing in a flash. Roblox also has error outputs and a command bar that can run in live games, which helps with debugging. All raw Studio sessions begin with systems for players, rigs, animation, movement controls, lighting, multiplayer, and some UI features. Roblox Studio provides the tools to tinker with any of these, alongside some external tool and software support. Your Roblox game can enhance any of these defaults or move away from them.

Ready to Be Modified

Roblox Lua enables the manipulation of already existing properties. Properties determine how objects look and function. These properties are prevalent in many class objects. For example, properties include the shape, color, or even material of an object.

Take this primitive shape, known as a Part, for example. In Figure 1.9, you can see its properties, including the color as medium stone grey and the material as plastic.

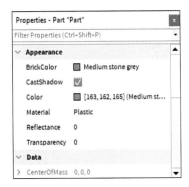

FIGURE 1.9
Some properties of a part as shown in Roblox Studio.

Elements of these properties that aren't grayed out can be modified through code or by the user. You can alter the properties of not only 3D objects but of anything else that you can place and use in Studio, such as particle emitters or user interface frames. Having the know-how about what properties can be used on Roblox will help you open your mind to more complex games. We talk much more about parts and properties, beginning with their introduction in Hour 2, "Using Studio."

Conceptualize with Ease

You can find free assets to help you get started in the Toolbox. They are streamed and then loaded in real time, so you don't need to install them. You can then combine them with manipulatable primitives, such as texturable blocks and spheres, to test new concepts or play around (Figure 1.10).

If working with just blocks isn't your style, you can sculpt levels with Roblox's Terrain Editor (Figure 1.11), which optionally includes foliage.

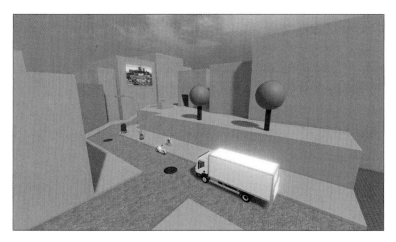

FIGURE 1.10
A blockout test environment from using parts, free assets, and modified property settings.

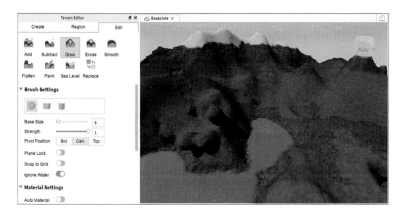

FIGURE 1.11
Using the Terrain Editor in Roblox Studio to sculpt an environment.

Roblox has signed a license agreement with APM Music that provides thousands of audio tracks from APM to be played only in games with no worries of copyright claims. If you need a certain soundtrack, search the audio from APM uploaded by Roblox to find what you need.

Plugins

Roblox Studio supports the ability to install user-created plugins to enhance your experience in developing. Developers have made their own tools, their own content installers, or external software support interfaces. Some plugins will allow you to generate trees, fill gaps when it might be otherwise tedious to do so, scan for viruses, and even edit light inside of objects. Roblox's official plugins include a language translation software, and animation and rigging tool for characters.

No Wait Times for Publishing or Updating

Roblox is hands-free when players want to schedule something new for their games. Updating a game doesn't require contacting a middleman or retailer. Every Roblox game has a Configuration page for both Places and Games.

In games received from regular online retailers, it's expected that users will install updates when they are released. Not on Roblox! Because assets are streamed, like during a Roblox Studio session, updates also will be streamed when players start joining and loading up your game. This does not extend to the Roblox client itself because new software updates to the platform occasionally install onto your device.

What's Inside Roblox's Engine

Roblox's engine, Roblox Studio, provides a lot of components that other engines would have the player make themselves. This makes developing on Roblox less expensive in terms of time and resources. As a designer, you can share your projects so users can experience them quickly with little hassle.

Networking

Roblox is responsible for server hosting and has already provided online connectivity services for you. Getting your game online is easy because you don't have to tinker with hardware or software outside of Roblox. Server hosting is automatically established as soon as games are published. Hosting can be private or public and can range from playing solo to playing with 100 players maximum. The number of players is set by the developer through the website. For well-balanced performance, a player cap of 20 or 30 is practical for an action-packed environment.

Roblox Lua supports web services that will help connect Roblox games with real data from the Internet. HTTP Service can be used to connect a Roblox game with third-party services and can provide things such as analytical data. Another form of this is called Asset Service, which can load asset data from the Roblox website, like a catalog item's description or creator name, into a live game.

Roblox also provides developers with a security measure called Filtering Enabled, which forms a client-server structure for games and prevents clients from replicating things onto the server, thus reducing the abilities of exploiters and hackers. Adding additional game security is up to the developer.

Physics

Roblox has its own physics simulation engine so your environments and assets can be as dynamic as ever. Every 3D object in Roblox can have physics with toggleable collisions. Meshes automatically generate their collision meshes upon loading but can be limited to their hull or a bounding box for performance purposes. To disable simulated physics on something, it must be Anchored.

Constraints and Attachments are provided in the Roblox engine, such as ropes, springs, welds, and more (Figure 1.12). These can result in some complex contraptions with coding. Vehicles, hydraulics, and suspension systems are all possible with constraints. You can put them on parts and other 3D objects to help gain better control of physics (Figure 1.13).

FIGURE 1.12
An explosion affecting an environment built out of welded parts, accompanied by custom UI.

FIGURE 1.13
Players build boats using materials of different density to find out what floats. *Build a Boat for Treasure* by Chillz Studios.

Rendering

Visual fidelity on Roblox can support all sorts of environments for any game. Lighting shaders on Roblox support atmospheric fog, particles, real-time lighting, shadow maps, ambient occlusion, anti-alias, and various screen effects (Figure 1.14). Roblox also has the capability for physically based rendering, ready for shaders such as normal maps and metal/roughness.

All these systems work with predetermined sets of graphics options on a scale of 1 to 10. Roblox allows users to set graphics to automatically change these levels higher or lower based on a user's performance. Additionally, games can be streamed, where players will be loaded in instantly, at the cost of letting things load as the player encounters them in the game.

FIGURE 1.14
Recreation of the Roblox headquarters.

NOTE

Developers Design the World

The look and feel of a game is completely up to you. You can even change the provided user interface and player avatars. Although Roblox is often portrayed as a 3D universe, 2D experiences are possible as well.

Cross-Platform Support

Roblox is supported on multiple devices with cross-platform support. That means a user on a tablet can meet with a user on a different device, such as a console, in the same game! Developers can prepare their Roblox games to be run on any of the following devices:

- ▶ macOS computers

- ▶ Windows PCs

- ▶ iOS and Android devices

- ▶ Xbox One

- ▶ Virtual reality headsets

Roblox originated as exclusive to desktop computers, and trends and core features were built for such. To make games built to run outside of the desktop spectrum, you must account for the different specifications of your preferred devices, such as user-interface scaling and input. Studio supports device simulation, so it's possible to test games in a cross-platform environment before releasing them (Figure 1.15).

FIGURE 1.15
Device simulation using a mobile phone interface.

Free, Free, Free

There's huge potential for a game developer who's able to access an already existing market with no server-maintenance costs, a game engine, your own social pages, and a generous offer of cloud-storage—all for free! This can put you at an advantage in time and cost compared to working with another engine. Remember that Roblox will not make you pay upfront to utilize its game development tools, such as Roblox Studio.

The most you'll have to interact with monetary hurdles on Roblox is usually for extending social and economic features such as deeper avatar customization and getting more Robux to participate on the site more. Additionally, entering the DevEx program to exchange your virtual currency for real currency requires a Premium subscription, which will in turn support the upkeep of Roblox.

Unlimited Possibilities

Roblox houses all sorts of game genres and projects. There is no definition on what a real Roblox game is. There are only Roblox experiences. You're not limited to common categories and you can even go design your own type of genre if you desire. Some trending games on Roblox include round-based minigames, open-world experiences, technical experiments, and art portfolio showcases (Figure 1.16).

FIGURE 1.16
Dungeon Delve by Roblox Resources.

Express Your Own Aesthetic

Outside of the Roblox brand and some default assets provided, Roblox does not promote a specific aesthetic. The look and feel of games is completely up to developers. As such, Roblox games employ a wide range of visuals, from very quirky and cartoon environments (Figure 1.17) to very sophisticated realistic experiences (Figure 1.18). There's something for everyone.

FIGURE 1.17
In *World // Zero* by Red Manta, bright colors and custom characters come together in a cartoony fantasy MMO.

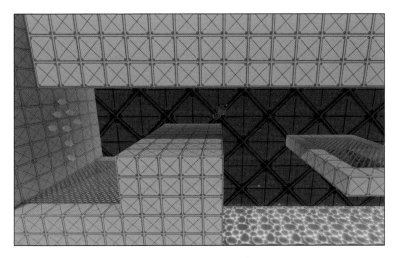

FIGURE 1.18
Jump, explore, and collect candies in the retro world of *Robot 64* by zKevin.

Summary

In this hour, you've learned about the culture and features that make Roblox a standout platform. Being able to communicate using the site's social features and understanding how to use what Roblox has provided for you is a great first step toward having a presence in the development community. Remember that every developer has started somewhere, and knowledge about the platform is a great start in taking advantage of more technical topics.

Q&A

Q. Can I advertise my Roblox career outside of Roblox?

A. Yes, you can use other websites to enhance and grow your audience as long as you don't violate the Roblox rules and Terms of Service (https://en.help.roblox.com/hc/en-us/articles/115004647846-Roblox-Terms-of-Use).

Q. How is copyright handled on Roblox?

A. All expected copyright laws apply on Roblox. Using intellectual properties that were made by someone else can be grounds for moderator action, especially if you have monetized those things. This extends to content made outside of Roblox. If certain copyright permissions are granted, you may use it to the said extent in your work if obtained in that period.

Q. Can I ask Roblox to make a feature for my game?

A. Roblox will remain hands-free on the development of content. If you want to suggest features for the platform as a whole, you may do so in the Developer Forums.

Workshop

Now that you have finished, let's review what you've learned. Take a moment to answer the following questions.

Quiz

1. How do you join the Developer Forum?

2. What technology is used to store Roblox's assets?

3. True or False: Playing and developing on Roblox requires only one account.

4. Aside from a game engine platform, Roblox can be associated as a(n) _____ platform.

5. True or False: I will need to pay to give an uploaded asset its own page.

Answers

1. Regularly browse and read the various content in the Developer Forum. After you've spent enough time browsing and reading the resources, you will automatically level up (from visitor to member and even to community sage) and can start making your own posts.

2. Cloud storage.

3. True. Developers have the same accounts as players.

4. Social.

5. False. All creations and assets automatically get their own page when uploaded.

Exercises

Follow the exercise below to make a Roblox account on a desktop computer. You only need one account to start developing and playing.

1. On a web browser, go to https://www.Roblox.com.

2. If prompted, apply your correct birthday by clicking on the month, date, and year drop-down buttons.

3. Create a unique username. This will be your online name. It must be 3 to 20 characters (letters and numbers). Do not reference anything that could compromise your privacy, such as your real name.

4. Create a password that only you can memorize. It must be at least eight characters long.

5. If prompted, select your preferred gender. This will be used to give you free items associated with your preference on Roblox.

6. Read the Terms of Use and Privacy Policy to understand them.

7. Click Sign Up to be able to explore and use Roblox. Welcome to your home page!

Follow the bonus exercise to add some security to your Roblox account and take your first step to personalize it a bit. You will need an email address, your guardian's permission if you're younger than 13, and an Internet connection.

1. On the top navigation bar, click the gear icon, and click Settings in the drop-down menu.

2. Select the Account Info tab.

3. Select the option to add your email address or your parent's email.

4. Enter your email address or ask a guardian to add their email address if such permission is needed. Verify your credentials and enter your email.

5. Roblox sends a verification email to the address you entered. Gain access to the email account and verify your Roblox account.

6. Return to your Settings page.

7. Add a description to appear on your profile page. Don't add information that will risk your privacy. Get creative!

HOUR 2
Using Studio

What You'll Learn in This Hour:

▶ How to install and launch Roblox Studio
▶ How to use Studio templates
▶ How to navigate game editor
▶ How to create a part
▶ How to translate, scale, and orient parts
▶ How to save and publish your project
▶ Playtesting

Now that we've explored the culture and features that make Roblox special, you can start to unleash your creativity with Roblox's free game engine, Roblox Studio. Roblox Studio is a playground for developers to create, share, and play their games on the Roblox website. What's great about this platform is that you can easily build everything from volcanic islands to urban cityscapes and then drop a character into that world to immediately start playing. Imagine a huge playground filled with all the tools you need to build imaginary worlds—that's Roblox Studio.

In this hour, you'll learn how to install Studio, and then you'll learn how to use Roblox Studio with the help of templates. You'll also learn how to arrange your workspace to hold objects in the 3D world, the difference between saving and publishing your project, and finally how to test your game before publishing it to the public.

Installing Roblox Studio

We've explained how Roblox Studio is a free and immersive platform for game developers to build different terrains, cities, buildings, race games, and much more. You don't need years of coding experience or a degree to make fun games; all you need is your imagination and hands-on learning in the Roblox Studio. Roblox Studio is extremely intuitive to use. Because Roblox is cross-platform, developers can install Studio on both Windows and Mac systems.

Use the following steps to install Studio:

1. Go to https://www.roblox.com/create.

2. Click Start Creating and then click the Download Studio button in the pop-up window.

3. Navigate to the folder where you have downloaded Studio and double-click the file to install it.

NOTE

System Requirements

For Roblox Studio to run efficiently, there are some OS/hardware specifications:

▶ Roblox Studio cannot run on Linux, Chromebooks, or mobile devices such as smartphones.

▶ A Windows computer with at least Windows 7 installed, or a MacBook with version macOS10.10.

▶ A minimum of 1 GB of system memory.

▶ Internet access to download Studio and updates. It also lets you save projects (publish) to your Roblox account.

For an enhanced Studio experience, you should also have these things (not mandatory):

▶ A mouse with a scroll wheel, preferably a three-button mouse.

▶ A video card that's dedicated and not an integrated card.

Troubleshooting the Installation

If you've followed the necessary steps to install Studio but you're experiencing installation conflicts, there are a few things you can do to troubleshoot the errors:

▶ If you've added new hardware or drivers recently, remove and replace the hardware to determine if it's causing the problem.

▶ Run diagnostics software and check information on troubleshooting the operating system.

▶ Restart the computer.

▶ Uninstall and delete all the Roblox files and reinstall the latest Studio again, if required.

If you are still finding errors, you can also reference the Roblox Support forums online for additional tips.

Opening Roblox Studio

Once you are done installing the Roblox Studio, you need to open it:

1. Double-click the desktop icon if you are on Windows or click the Dock icon if you are on a Mac to open a login window (Figure 2.1).

2. Enter your Roblox username and password.

3. Click the Log In button.

FIGURE 2.1
Roblox Studio login window.

Once you are logged in, you see a page with different templates and a menu sidebar with New, My Games, Recent, and Archive (Figure 2.2).

The following sections provide a quick introduction to these templates and the rest of Studio; then you can begin experimenting with the utilities of Studio.

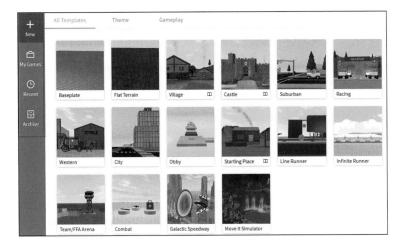

FIGURE 2.2
Roblox Studio home screen.

Using Studio Templates

When you first open Roblox Studio, under New, you see three tabs: All Templates, Theme, and Gameplay. Templates are prebuilt projects, and you can use them as a guide to build your own game world.

All Templates

The All Templates tab (Figure 2.3) is a combination of the Theme and Gameplay tabs. You can use these templates as a start for your games. For example, if you're building a medieval game, the Castle theme is equipped with feudal details, or if you want to build an interactive obby, you can build off the Obby gameplay template. Two simple templates are a good place to start:

▸ **Baseplate:** This is a popular choice to start with. The baseplate itself is easy to delete, leaving a blank canvas to work with.

▸ **Flat Terrain:** Has a flat plane of grass terrain instead of a baseplate. You can modify or clear the terrain using the terrain editor.

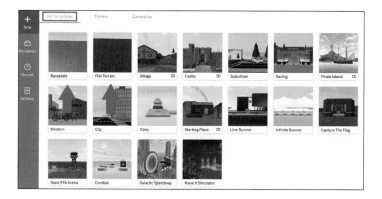

FIGURE 2.3
Roblox Studio home screen lists various templates available, such as simple templates Baseplate and Flat Terrain.

Themes

Themes are a combination of gameplays and more, and together they make a new world. It sets a mood for your game—for example, a space combat game will have asteroids and other galactic components. Roblox provides some prebuilt themes that are ready to use and modify however you would like. As you explore the game world, descriptions point out its use case or features, including tips on how the effects were created in case you want to re-create them yourself.

An example of a prebuilt theme is Village (Figure 2.4). You can explore the houses in the village and move along the pathway through the town, which leads you to a river, a bridge, and finally the dock, across which you can see small islands.

FIGURE 2.4
Example of a prebuilt *Village Theme* available in Studio.

Gameplay

Some templates include interactive gameplay. For example, this can include Team Deathmatch, Control Points, Capture the Flag (Figure 2.5), and more. A great thing about these templates is that developers can take them apart and extract any specific facet that they want—for example, using in-game radar or team spawn points. These templates help with components such as what a player can do in a game, what the goals are, and how a game can be modified.

FIGURE 2.5
Example of a prebuilt *Capture the Flag* gameplay template.

Working with the Game Editor

Now that we've familiarized ourselves with Studio's homepage, let's click on the Baseplate template to get started. This opens the game editor (Figure 2.6).

The game editor is, as the name suggests, a place where you can create, modify, or test your game. At the top of the game editor, you see different tabs on the menu bar (Figure 2.7).

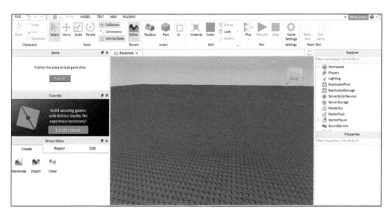

FIGURE 2.6
The game editor enables you to create, modify, or test your game.

FIGURE 2.7
Roblox Studio menu bar.

▶ **Home tab:** A concise tab of all the features that are frequently used. These features are on the Home tab for easy access.

▶ **Model tab:** Has more building tools apart from move, scale, and rotate. It's also where you can create spawn locations and special effects such as fire and smoke.

▶ **Test tab:** Helps for testing your game. There are two options underneath: Run and Play. Run will run a simulation of what will happen to the bricks and surrounding elements, and Play will let you play your game.

▶ **View tab:** Lets you toggle the different windows available in the Roblox Studio. If you need to use a window that is closed, you can find them under the View tab.

  ▶ The main windows are Explorer and Properties, which are discussed detail in later in this section.

  ▶ The Actions section has several display features. You can take screenshots or record videos here and also toggle between full screen and windowed views.

▶ **Plugins tab:** An add-on to Studio. These are generally not included by default. Plugins add new custom behavior and features. You can either install plugins made by the Roblox community or create your own plugins.

Below the menu bar is a ribbon bar (Figure 2.8). The tool options change as you move between menu bar tabs.

FIGURE 2.8
Roblox Studio ribbon bar.

In the following sections, we explain some of the editor's basic features and most frequently used features and discuss how to prepare your project for publishing on Roblox.

Arranging the Game Editor Workspace

Since this is the first time you are opening the game editor, extra windows that you don't require right now will automatically open on the left side. To organize the workspace in an optimal way, close the extra windows so you have more space to create.

By default, the Explorer and the Properties windows will be open (Figure 2.9), aligned one beneath the other on the right side.

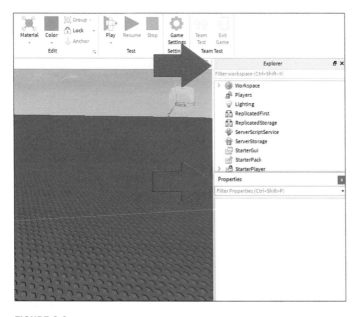

FIGURE 2.9
Workspace arrangement with the Explorer and Properties windows one below the other.

NOTE

Some Features of the Game Editor Workspace

The next time you relaunch Roblox Studio, your workspace arrangement remains intact. It is a one-time fix, unless you undo your arrangement.

When the Property window undocks, it gets difficult to dock it back below the Explorer window. It either docks itself aside or over the Explorer window. To fix this, undock both the windows and close them. Go to the View tab, open the Explorer window, dock it on the right-hand side, and then close it. Do the same with the Properties window and close it. After all this, reopen the Explorer and then the Properties window. This will align them one above the other.

Working with the Explorer Window

The Explorer window is the hierarchical representation of all the objects used in your game. It is the most crucial window because it lists all the organizing, viewing, and testing features of a Roblox game.

It uses the concept of parenting to organize all the objects. The object Game is hidden at the top of the hierarchy. For example, in Figure 2.10, you can see Workspace parent has the following children nested underneath: Camera, Terrain, and Baseplate.

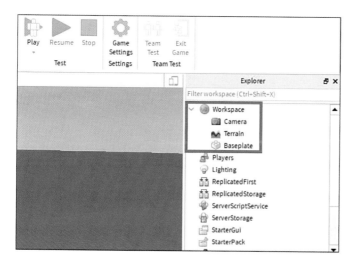

FIGURE 2.10
Objects nested under Workspace in the Explorer window.

If you want to create more child objects, you can hover over Workspace and click the plus symbol to the right (Figure 2.11). This will list all the objects that you can create. You can also drag and drop it into the desired parent object.

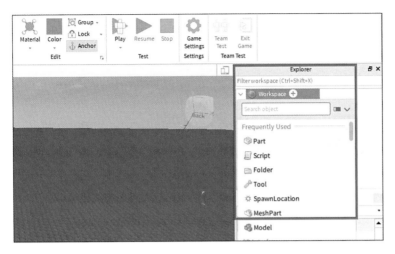

FIGURE 2.11
Add more children to your Workspace.

One of the most important children you will work with is a part, which is the foundational building block of Roblox. These physical 3D objects are also known as bricks, and when they are in the Workspace, they can interact with each other.

Creating a Part

To create a part, from the Home tab, navigate to the Insert menu in the ribbon bar and click Part (Figure 2.12).

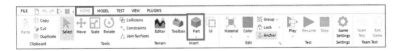

FIGURE 2.12
Create a part.

A part will appear at the exact center of your camera view (Figure 2.13). Use the **camera controls** shown in Figure 2.14 to move your camera, rotate the view, and zoom in and out.

FIGURE 2.13
Part appears in your baseplate and in your Explorer.

Control	Action
W A S D	Move the camera
E	Raise camera up
Q	Lower camera down
Shift	Move camera slower
Right Mouse Button (hold and drag mouse)	Turn camera
Mouse Scroll Wheel	Zoom camera in or out
F	Focus on selected object

FIGURE 2.14
Camera controls.

To give your new part a name, do the following:

1. Double-click the part in your Explorer window.

2. Rename the part. Roblox convention is for parts to be named in PascalCase, which means the first letter is capitalized—for example, EndZone or RedBrick.

Note that your name can contain spaces, but we won't use spaces at this point in case we want to be able to access the part via code later.

You can use the Explorer to select and work with parts even if you can't see them in the game editor window.

Working with the Properties Window

When you add a part to your Workspace, you'll notice the Properties window (Figure 2.15) fills with information.

FIGURE 2.15
The Properties window lists all the details about the newly added part.

Like any object, a part has properties such as size and color, and the Properties window shows all these details about how an object looks and behaves. In the next chapter, we'll go into further detail about properties of a part and how you can manipulate them.

Translating, Scaling, and Orienting Objects

You've learned how to create a part; now you can make it move! In Roblox Studio, it is possible to move (translate) and rotate (orient) objects in the scene. There are multiple ways to get the same results, but in this section, we will solely use the Roblox Studio default tools and keyboard shortcuts.

There are two settings you can use to get greater control when moving parts: snapping and collisions.

▶ **Snapping** is the amount a part will move, scale, or rotate at a time. Snapping is useful when creating items that need to be exactly aligned, like how walls of buildings need to be at 90-degree angles.

▶ **Collisions** happen when two objects (or rigid bodies) intersect or get within a certain range of each other.

Because these two settings are most used when playing with two or more parts, turn them off for now while you freely move a single part around. Later, you'll turn them back on when we discuss how they work.

▶ **To turn OFF snap:** In the Model tab, uncheck the box next to Rotate or Move (Figure 2.16).

FIGURE 2.16
Turn off snap.

▶ **To turn OFF collisions:** In the Model tab, collisions are on if the button is highlighted gray. Click the Collisions button to toggle it off (Figure 2.17).

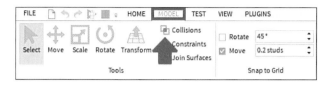

FIGURE 2.17
Turn off collisions.

Translating

Now you can freely start translating, or moving, objects. Go to the Model or Home tab and click the Move icon (Figure 2.18).

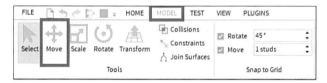

FIGURE 2.18
Move tool.

Now, a gizmo should appear on the selected objects. When you click, hold, and drag one of the arrows, the object moves along that axis (Figure 2.19).

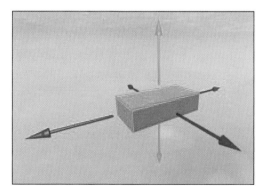

FIGURE 2.19
Moving the gizmo.

Scaling

To **scale** objects, go to the Model or Home tab and click the Scale icon (Figure 2.20).

FIGURE 2.20
Scale tool.

The gizmo should appear again, this time with orbs on selected objects. When you click, hold, and drag one of the orbs, the object scales along that axis (Figure 2.21).

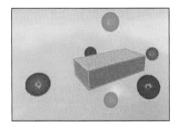

FIGURE 2.21
Scaling a gizmo.

If you want to scale on two sides simultaneously, hold Ctrl (Windows) or Command (Mac) while clicking, holding, and dragging one of the orbs.

If you want to scale while keeping the current proportions, you can do so by holding Shift while scaling.

Rotating

To rotate objects, go to the Model or Home tab, and click the Rotate icon (Figure 2.22).

FIGURE 2.22
Rotate tool.

Another gizmo should appear, now with orbs and circular, connecting lines on selected objects (Figure 2.23). When you click, hold, and drag one of the orbs, the object will rotate along that axis.

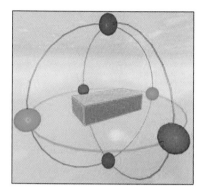

FIGURE 2.23
Rotating a gizmo.

Transforming

The transform tool (Figure 2.24) is particularly important as an all-in-one building tool. It enables multiple moves, scales, and rotations within one continuous operation. Think of it as a bundle of move, scale, and rotate. Basically, it can transform your part in any way possible. It also can lock an axis and snap to the grid.

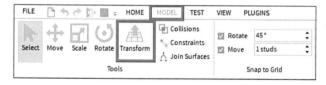

FIGURE 2.24
Transform tool.

With your part selected, click on the transform tool and markers for manipulation appear around your part (Figure 2.25).

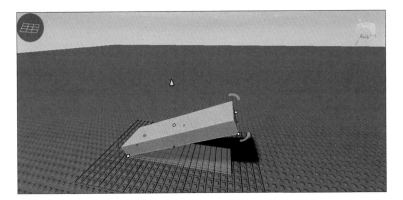

FIGURE 2.25
Using the Transform tool.

▶ The yellow cone is used to move the part on different planes on the Y axis. We can drag
the part on its own plane once the plane is set.

▶ The red, green, and blue arcs are used to rotate the part by 360 degrees on the X, Y, and Z
axes.

▶ The white boxes are used to scale the side of the part to which they are attached. The scaling happens in the measurement of studs, which is the measurement of each single square
that forms the baseplate.

Snapping

Now that we understand the basics of moving a single part, let's revisit snapping and collisions.
As a reminder, snapping is the amount a part will move, scale, or rotate at a time, and it allows
you to align an object perfectly. There are two types of snapping: Rotation or Move.

▶ **Rotation** snapping enables you to turn an object by the given number of degrees. In this
case, all objects will rotate 45 degrees each step.

▶ **Move** snapping counts for both moving and scaling. In this case, any object moves for one
stud each step. Objects scale one stud each step.

Keep in mind that when you scale from the center of an object, it will scale one stud on both
sides. It will then equal two studs total.

To turn snap back on, you will check the box next to Rotate or Move in the Model tab. Then, in
the Rotate or Move fields, you can adjust your setting by the number of studs you want to move
(Figure 2.26).

FIGURE 2.26
Snapping options.

Collisions

You can turn collisions back on and notice how they affect movement. In Roblox Studio, the collisions feature lets you control whether parts can move through each other. When collisions are on, you can't move a part into any place where it overlaps another part.

To turn collisions back on, click the Collisions button in the Model tab. This toggles it on and highlights it gray (Figure 2.27).

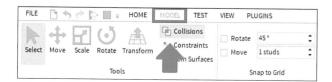

FIGURE 2.27
Collisions on.

Now as you move parts, you may notice a white outline whenever a part touches another part. This indicates that a collision is happening. We'll talk more about collisions in later hours.

Anchoring

We've talked a lot about making parts move in this chapter, but what if you don't want a part to move? If you want a part to be immobile, you need to anchor it. When you anchor a part, it remains static even when you're playing the game and other players and objects run into it. To anchor a part, do the following:

1. Go to the Properties window.

2. Scroll down to Behavior.

3. Check Anchored (Figure 2.28).

FIGURE 2.28
Anchoring a part.

You can also easily Anchor and Unanchor parts with the Anchor button located in the bottom of Model tab or Home tab (Figure 2.29).

FIGURE 2.29
Anchor button.

▼ TRY IT YOURSELF

Anchoring Parts

To practice anchoring parts, do the following:

1. Create a part.

2. Move it left.

3. Rotate it 90 degrees with Snap to Grid.

4. Check the Properties window to see if it's anchored.

Saving and Publishing Your Project

Now that you are creating in the game editor, you will want to save your progress on projects from time to time because you don't want to lose any of the work you've accomplished. When you're ready for people to enjoy your creation, you may also want to publish it.

Saving Your Project

Roblox doesn't autosave your projects for you, so you need to save them. There are two places where you can save projects:

▶ **On your local desktop:** On the game editor menu bar, click File at the top-left corner, and then click Save to File. This retains the template name and saves the project as an .rbxl file. Instead, if you choose the Save to File As option in the same drop-down menu, you can rename the file (Figure 2.30).

▶ **On the Roblox server:** You can also save your project on the Roblox server by using the Save to Roblox As option in the same drop-down menu. This saves your work to a secure place in the Roblox Server but does not make it accessible to the public.

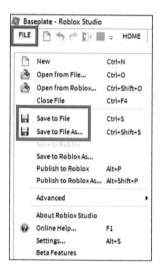

FIGURE 2.30
The Save to File commands are under the File option.

Publishing Your Project

What's the use of creating a game if no one can play it? To make it public and monetizable, we need to Publish the project by choosing the option Publish to Roblox. Publishing makes your game public and allows other players on Roblox to play it. Following are the steps to Publish to Roblox:

1. Select File, Publish to Roblox to open the publishing window.

2. Enter a name and an optional description.

3. When ready, click the Create button.

Reopening Your Project

When you want to reopen the project you were working on, you can find it on the Studio home screen (Figure 2.31) as follows:

▶ **File:** Select File, Open.

▶ **My Games:** If you have published your game to Roblox, your game will be in My Games.

▶ **Recent:** Look in Recent for all files that you've recently had open.

FIGURE 2.31
Reopen previous projects from the Studio home screen.

Playtesting

Playtesting is the process of playing the game to make sure everything works and figuring out how to make it even better. Don't skip this step because it's critical for a successful game. It's good practice to playtest your game whenever you make changes. You should also test your game in various modes. You can make changes in Play mode, but those changes won't be saved. You'll have to do them again when you go back to editing.

TIP

Playtesting Practices

When you playtest, do the following:

- ▶ Make sure your game works, particularly changes you just made.
- ▶ Look for areas that can be improved.
- ▶ When you are exploring or playtesting templates, make sure you thoroughly look at how the parts are named and grouped together.

Playtesting Your Game

To playtest your game, follow these steps:

1. Save your game. Don't forget to change the filename.

2. Press the Play button in the top menu bar. You can also find the Play button in the Home tab under the Test menu (Figure 2.32).

FIGURE 2.32
The Play button for playtesting your game.

Stopping Playtesting

To stop playtesting, press the red Stop button either in the top menu bar or under the Test menu (Figure 2.33). Stop the Playtest before making changes. Again, the reason for this is because you can't save changes in lay mode.

FIGURE 2.33
The Stop button for playtesting your game.

▼ TRY IT YOURSELF

Playtesting Practice

Playtest the following two templates:

▶ Village

▶ Obby

Before playtesting, you can modify the places where the parts are placed. You can drag and drop parts and watch how their properties change in the Properties window, and you can modify materials or delete them. Don't forget to save it or publish the template under a new name, and if you try to add parts or effects, make sure they are not in playtest mode.

Summary

In this hour, you've seen how easy it is to use Roblox Studio to create and share games with millions of players. You learned how to install and use the Roblox Studio, as well as how to arrange the workspace, make changes to your template, and save and publish games on Roblox to share them with the public. You also learned how to playtest your changes to ensure the success of your game.

Q&A

Q. What needs to be done if Studio isn't installing?

A. Make sure your system has the minimum system requirements. If it doesn't and Studio still ends up installing, there might be problems running Studio.

Q. Can I modify a template?

A. These templates are prebuilt projects you can use as a start for your own games.

Q. Can I save changes made during playtesting?

A. Changes made in Play mode won't be saved. You'll have to do them again when you go back to editing.

Workshop

Now that you have finished this hour, let's review what you've learned. Take a moment to answer the following questions.

Quiz

1. How do you organize your workspace?

2. Which two common starting point templates can be developed from scratch?

3. How do you move your avatar around during playtesting?

4. True or False: Publishing your project on Roblox makes it visible to everyone.

5. True or False: The Transform tool is an all-in-one building tool.

Answers

1. Closing the extra windows will give you more space to see what you're doing and keep the Explorer and Properties windows aligned below each other.

2. Baseplate and Flat Terrain are two commonly used templates on which a game developer can develop from an entire game world.

3. We use the WASD or the arrow keys to move around.

4. True. Publishing saves your work to a secure place and allows other players on Roblox to play your game. (To make it public to everyone, go to "Game Settings" after the initial publish.)

5. True. The transform tool is an all-in-one building tool. It moves, scales, and rotates in a precise way.

Exercises

Follow the exercise below to gain additional insight into the Roblox Studio.

1. Open a new Baseplate template.

2. From the Home tab, add a part block.

3. Find the new part added to the Explorer window under Workspace. Rename it as `CenterPart`.

4. Rename and save your baseplate; then publish it to Roblox.

5. Playtest your game.

This second exercise combines a number of things you've learned the last two hours. If you get stuck, don't forget to refer to the previous pages in this chapter! You're going to make a simple obstacle course (commonly referred to as an "obby" in Roblox).

1. Start with a couple of parts. Make sure that Anchored is enabled and place them in the sky. Feel free to color them any color that you want, or even add decals or textures!

2. Add another part at one end of the parts. This will be the start of your obstacle course game. Make sure that it is also Anchored.

3. Add your final part at the other end of the parts. This will be the end of your obstacle course game. Make sure that it too is anchored.

4. Playtest your game. Test out your game by flying over your starting point, clicking the blue arrow underneath Play in the Home tab, and choosing Play Here.

5. Bonus: Add a Spawn object from the Gameplay section of the Model tab at the top of Roblox Studio to avoid having to press Play Here and having all players start at the beginning. (It is anchored by default!)

TIP

Keep these tips in mind.

► Add at least five or six parts of differing sizes and shapes to create a jumping puzzle for players. The beginning jumps should be easier than later jumps.

► Playtest your game throughout the creation process to make sure you can make each jump and that all parts are anchored.

HOUR 3
Building with Parts

What You'll Learn in This Hour:

▶ How to change a part's appearance
▶ How to create decals and textures

In the last hour, you learned how to use Roblox Studio to create unique games to share with other players. This hour explains parts in greater detail and describes how they can be used. Parts can be big or small, and they can have different textures and colors. Like everything in Roblox, parts are limited only by your imagination for building all sorts of creations. You can use parts to create props, cityscapes, and even vehicles. In this hour, you learn how to create a part, manipulate the look of a part, and create, add, and modify decals and textures.

Creating a Part

Let's review how to create a part:

1. Go to All Templates, Baseplate.

2. From the Home tab, click Part on the ribbon bar (Figure 3.1) to spawn a part at the center of your camera view.

FIGURE 3.1
Create a part.

Changing a Part's Appearance

Once you have created a part, you may want to modify how it looks. You do this by changing the part's properties in the Properties window:

1. Select the part in your workspace.

2. Go to the Properties window to see all the properties of the selected part and scroll down to the appearance tab (Figure 3.2).

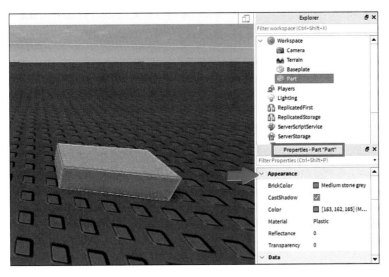

FIGURE 3.2
Part properties on the Appearance tab.

As you can see, there are multiple properties for the appearance, including Color, Material, Reflectance, and Transparency. The following sections go into more detail about different appearance properties.

TIP

Show/Hide Windows

If windows are hidden or missing, you can open them by going to the View tab in the top menu bar (Figure 3.3). Click the windows you need throughout this tutorial.

FIGURE 3.3
Use the View tab to show/hide windows as needed.

Color

Color is used to change the color of the part's surface. Click the check box next to Color in the Appearance tab to open a color picker (Figure 3.4), where you can select any shade for your part.

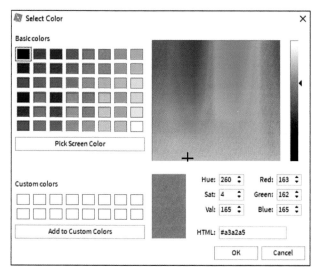

FIGURE 3.4
The color picker.

Material

You can use materials to add more detail to your part to provide a more realistic appearance. As in the real world, changing a part's material also affects the density and behavior of the part—for example, Marble is more dense than Grass. The default material is Plastic, but you can change materials by clicking the Material drop-down menu and selecting a different option (Figure 3.5). With these materials, you can create anything from a dense forest to an urban concrete jungle.

FIGURE 3.5
The Material drop-down menu.

There are a total of 22 materials currently in Roblox, which are shown in Figure 3.6.

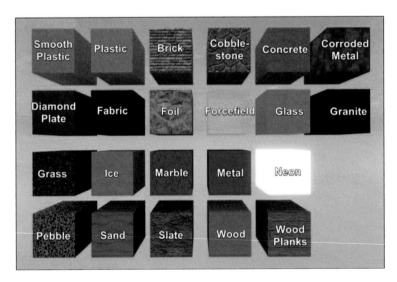

FIGURE 3.6
All currently available materials.

Reflectance and Transparency

The remaining properties are Reflectance and Transparency. Reflectance adds glossiness to a part's surface. When Reflectance is set to 1, the surface is completely glossy; when it is set to 0, it is completely dull (Figure 3.7).

FIGURE 3.7
Reflectance examples.

Transparency makes the part completely see-through when set to 1 and completely visible when set to 0. This is helpful when you're building glass surfaces and other transparent objects. Figure 3.8 shows some examples.

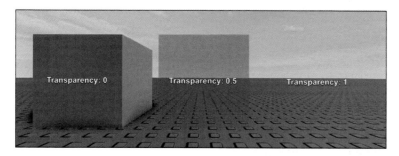

FIGURE 3.8
Transparency examples.

▼ TRY IT YOURSELF

Change a Part's Color, Texture, and Shape

Now it is time to use all the knowledge you have learned through this tutorial so far. Try to create the yellow cobblestone part shown in Figure 3.9.

FIGURE 3.9
Yellow cobblestone sphere part.

Creating Decals and Textures

Materials are great and allow for many creative options, but they are not the only method to add details to parts. You also can use decals and textures.

Decals are a different version of textures and have different uses. They stretch to fill the entire face of a part (Figure 3.10).

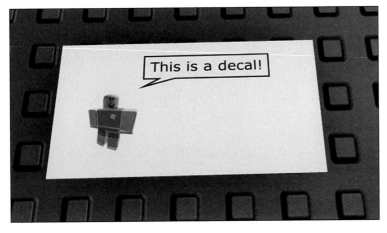

FIGURE 3.10
Example decal on a part.

Textures, on the other hand, have a few properties that allow for repetitive textures to be used side-by-side, among other things. In the zoomed-in detail for Figure 3.11, notice the uneven "seam" where the texture repeats, which is different than the whole image displayed on a decal, as in Figure 3.10.

FIGURE 3.11
Example texture on a part.

Decals

Decals are best used when you want the texture to stretch across an entire surface, such as billboard advertisements on the side of the road. You can create and upload a decal image to personalize your game objects. Follow these steps to create and upload your own decals:

1. Make sure your game is published so that the upload option is available. For a reminder on how to publish your game, review the end of Hour 2.

2. If you haven't already done so, use an image editor program—such as Photoshop or GIMP—to create the decal image you want and save it.

3. In Game Explorer, click Import. (Note that you may need to enable Game Explorer in the View tab.)

4. Select your saved decal image and click Open. This will make your image available for use in your game.

Once your image has been uploaded, add a decal to your part and apply your new image to the decal by following these steps:

1. Hover over the part in Explorer and click the plus button.

2. Select Decal from the context menu (Figure 3.12). A yellow border appears around the part to show where the decal will be placed.

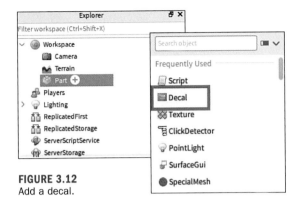

FIGURE 3.12
Add a decal.

3. In Properties, click the empty field next to Texture (Figure 3.13).

FIGURE 3.13
Decal property Texture.

4. Select your image from the drop-down menu. The image appears in the decal on your part.

This Texture property is reserved for the image source for the decal. The decal has other properties as well: Face, Color3, and Transparency.

▶ **Face** is the side, or face, of the part that is displaying the decal.

▶ **Color3** allows you to change the image's color. However, the Color3 value only adds color to the image, rather than overwriting it.

▶ **Transparency** allows you to make the image transparent (Figure 3.14), just as you made parts transparent earlier in this hour.

FIGURE 3.14
Decal transparency.

In Figure 3.13, you may have noticed that the Color3 value for the brick in the example is RGB [255, 255, 255], which is the RGB value for white. But the brick's actual color includes shades of yellow, black, and pink to name a few. If you want to change the color of the brick to blue, for example, you could enter the RGB for blue [0, 255, 255] in the Color3 property field, but the result will look like Figure 3.15.

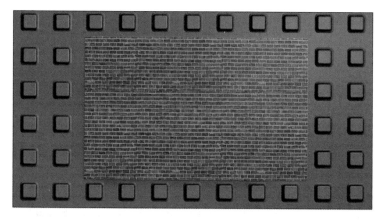

FIGURE 3.15
Colored texture.

Rather than the brick becoming blue, it puts a blue shade over the brick's original colors.

Textures

There are fundamental differences between textures and decals. One major difference is how the image is sized and positioned on the part's face. Textures can repeat over and over, so they're good for things like bricks or roads or when you need to be able to adjust a texture.

If you'd like to create and upload your own textures, the steps are the same as the steps for uploading a decal, except you select Texture from the context menu (Figure 3.16).

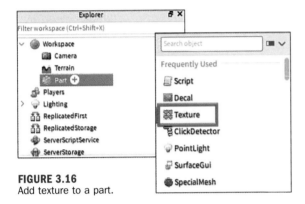

FIGURE 3.16
Add texture to a part.

Textures also have some of the same properties as decals (Texture, Color3, and Transparency), and they operate in the same way, too. However, textures have some additional properties—shown in Figure 3.17—including U and V, such as OffsetStudsU.

FIGURE 3.17
Texture properties.

Before we dive into them, though, you need to know what the U and V stand for. U is the 2D equivalent of the X (horizontal) axis and V is the 2D equivalent of the Y (vertical) axis. The reason why they are called U and V is because XYZ were already taken by the 3D axes.

OffsetStudsU and OffsetStudsV are basically the number of studs the image should be offset into the U and V directions. Figure 3.18 shows the same figure with the repeating texture that we used earlier in this hour.

FIGURE 3.18
Texture without offset. Notice the "seam."

As you can see, there is no offset whatsoever. But when we set the OffsetStudsU value to, say, 0.5, you can see in Figure 3.19 that it moves the image half a stud toward the side, making the seam less noticeable. It works the same way for V, but because our texture is so similar, it may be hard to tell. Try experimenting with different textures.

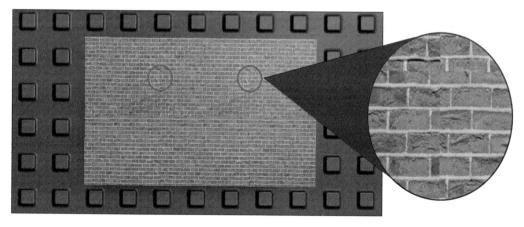

FIGURE 3.19
Texture with offset. Notice that the seam in the middle has been moved, and there are now two.

StudsPerTileU and StudsPerTileV are the image's size in studs into the U and V direction. If, for example, we set the StudsPerTileU value to 6, it means that the texture stretches over 6 studs per image. In Figure 3.20, notice how long the bricks are compared to the previous figures.

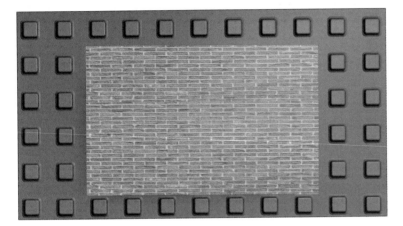

FIGURE 3.20
Texture with resizes.

Now if we change the StudsPerTileV value to 6, the result should be the nonstretched texture in Figure 3.21.

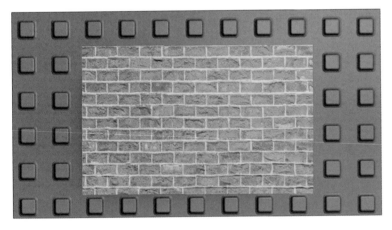

FIGURE 3.21
Texture with resizes.

TRY IT YOURSELF ▼

Create a Movie Theater

Now it is time to use all the knowledge you have learned through this tutorial so far. Try to create your own movie theater, complete with movie posters (Figure 3.22)!

FIGURE 3.22
Create your own movie theater.

TIP

It Doesn't Need to Look Exactly the Same!

Make sure to understand how to import and customize decals and use a variety of materials to add extra details to the place. We used a texture for the walls and a decal for the poster.

Summary

This hour built upon your knowledge of parts. You learned how to modify their appearance with decals and textures and explored real-world game applications of using them, such as building a brick wall for your game.

Q&A

Q. Can I modify a part's transparency?

A. Yes, using the Properties window, you can modify many of a part's properties, such as Color, Material, Reflectance, and Transparency.

Q. Can I modify a decal's properties?

A. Yes, decals and textures have their own properties, such as Color and Transparency, that you can also customize.

Q. Can I create and upload my own texture?

A. Yes, you can create any image you want and use it as a decal or texture in your game.

Workshop

Now that you have finished, review what you've learned. Take a moment to answer the following questions.

Quiz

1. True or False: Transparency makes my part completely see-through when set to 0.

2. True or False: The Color3 value allows you to overwrite color on an image.

3. True or False: Color and BrickColor are the same.

4. True or False: Scaling an object by 1 stud at the center will increase the size by 2 studs.

5. True or False: Decals stretch to the size of the face, whereas textures can repeat.

Answers

1. False. Transparency makes the part completely see-through when set to 1 and completely visible when set to 0.

2. False. Color3 allows you to change the image's color by adding color to the image, not by overwriting it.

3. False. This one is a bit tricky, but while both let you choose the color of a part, Color lets you choose any RGB value and BrickColor limits you to a palette.

4. True. When scaling from the center, increasing the scale will increase the size by double its increment.

5. True. Textures can repeat any number of times, whereas decals cannot.

Exercises

This exercise will combine a number of different things you've learned the last two hours. If you get stuck, don't forget to refer to the previous pages in this hour. Try to create a highway with a billboard on the side like the one shown in Figure 3.23.

FIGURE 3.23
A highway with a billboard posted next to it.

1. Begin with a road. Use multiple parts to make the main road, and separate parts for the middle divider lines.

2. Change the properties of the road and divider lines to use different materials and colors for the parts to make it look like asphalt and paint.

3. Create the base of the billboard with a cylinder part with a regular part on top. Change the materials of both.

4. On top of the billboard base, use another part and a decal to create a billboard.

5. **Bonus Challenge:** Instead of using multiple parts, create the road using one part and a texture instead.

HOUR 4
Building with Physics

What You'll Learn in This Hour:

▶ How to work with attachments and constraints

▶ How to work with the CanCollide property

▶ How to use hinges and springs

▶ How to use motors

Now that you've discovered how to manipulate 3D objects in Roblox, in this hour you will learn how physics are used to build a realistic interactive environment. If you want a working door or a moving fan in your world, you need physics.

The physics engine is responsible for determining how a part—whether it's a brick, wedge, sphere, or cylinder—moves in a Roblox game. The engine mimics real-life physics, making it easy to create physics-based contraptions, such as working cabinets or Rube Goldberg machines. In this engine, you set up motors, hinges, and springs and modify their properties to determine how fast or slow they should work in game.

In the following sections, we show two different ways you can create a door. The first method explains how to create a simple door that players can pass through but ignores Roblox physics. The second method takes advantage of the built-in physics features to create a door that opens and closes when players run into it.

In this hour, you use Studio to

▶ Build a door

▶ Move a player through the door by disabling CanCollide

▶ Re-enable CanCollide to make a more realistic door by adding hinges and springs

Working with Attachments and Constraints

Before you can build a door, you need to understand two key elements of mechanical construction: attachments and constraints. An attachment is where an object connects to a part. All attachments must be parented to a part, as shown in Figure 4.1.

FIGURE 4.1
Attachment points on a part.

A constraint connects two attachments. Constraints are elements, such as rods, motors, hinges, springs, and more, that can be used to build mechanical constructions. Figure 4.2 shows an example of a rod constraint.

In this example, you use a rod constraint with attachments to hang an unanchored part from an anchored part:

1. Create two floating parts, one above the other (Figure 4.3). Anchor the higher one, and don't anchor the lower.

FIGURE 4.2
Example image of a rod constraint.

FIGURE 4.3
Two parts that will be connected by a rod constraint.

2. Navigate to the Model tab and toggle on Constraint Details (Figure 4.4). This will make it easier to see the constraints and attachments you are creating.

FIGURE 4.4
Constraint Details button.

3. On the Model tab, select Rod Constraint from the Create button's drop-down menu.

4. Click the bottom of the top brick at the point where the rod attachment should connect, and then click the top of the lower brick. The rod constraint is created between two green attachments (Figure 4.5).

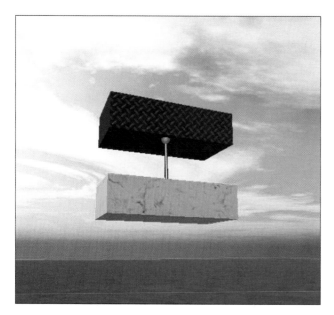

FIGURE 4.5
Two parts constrained together.

5. Playtest your game, and you'll see the unanchored part hanging from the anchored part.

Building a Door

Now that you have some practice with foundational attachments and constraints, you can try out two methods for building a door. For both methods, you need to create a simple door out of parts, using what you learned in the previous chapter.

1. Use three parts to create the door frame and a separate part for the door itself. Use Snap-to-Grid to keep everything lined up.

2. Anchor the door frame, but don't anchor the door itself.

3. Add a door handle to help you identify which side of the door is which (Figure 4.6).

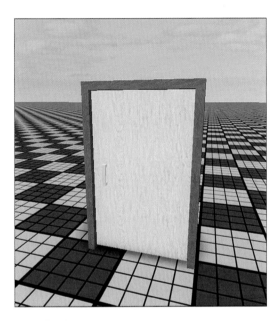

FIGURE 4.6
Door structure with handle.

To keep the door handle attached to the door as it moves, you need to use a weld, which is a type of constraint used to hold things together. Use the following steps to add a weld:

1. Click Create and select Weld from the drop-down menu (Figure 4.7).

2. Click the two parts that you want to weld together. You can only weld two parts with one constraint.

3. Continue to weld the other parts with the main part until the work is successfully done.

Figure 4.8 shows what parts are welded and which are anchored on a door to ensure the physics work properly. As you can see in this example, not only was the handle welded to the door, but the door itself is made of several parts welded together.

FIGURE 4.7
Triggering the weld option.

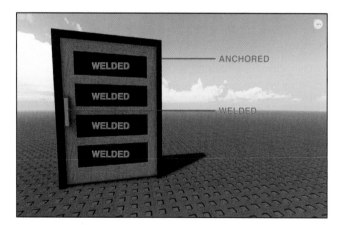

FIGURE 4.8
Welded parts versus anchored parts.

Now you have an object that looks like a door, but you need to make it function as a door and allow a player to move through it.

Disabling CanCollide to Move a Player Through the Door

One way to allow a player to move through the door is with a feature called CanCollide. The CanCollide property determines whether a part will physically collide with other parts or be able to go through other parts. The following are the two settings for CanCollide:

▶ When CanCollide is enabled for a part, players and other parts collide with it.

▶ When CanCollide is disabled, players and other parts can pass through the part.

In the example you're working with, disabling CanCollide on the door allows a player to move through the door. Follow these steps:

1. Select the door part.

2. Go to the Properties window, scroll to the Behavior option, and uncheck the CanCollide box (Figure 4.9).

FIGURE 4.9
Disabled CanCollide.

3. Playtest your game to see if you can walk through the door.

Now you have a door that a player or object can pass through. But, in the real world, a player wouldn't just walk *through* a door. They would open a door on its hinges, and it would spring closed behind them. Using Roblox's built-in physics features, you can add hinges and springs to the door to give it a realistic look and feel.

Adding Hinges and Springs

For players to interact with the door and push it open, you must turn CanCollide back on. Otherwise, players will just walk through the door instead of swinging it open. In this example of creating a realistic moving door, you need to use a hinge to swing it open and a spring to automatically close it.

HingeConstraint allows two attachments to rotate about one axis. You can use it for doors, cabinets, and more. HingeConstraint can even be used as a motor. It constrains the X axis of the two attachments so that they point in the same direction. The only way to change the rotation is to rotate the attachments themselves.

Opening the Door with Hinges

Use the following steps to add a HingeConstraint to create a realistic door:

1. Make sure CanCollide is enabled. (Refer to the previous section to find the CanCollide property and enable it.)

2. Move the door away from the door frame (Figure 4.10) to make it easier to place the attachments for the door hinges.

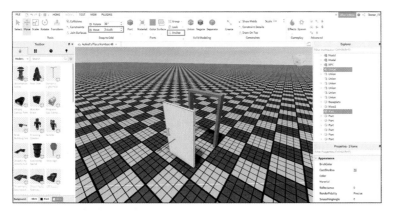

FIGURE 4.10
The door moved forward from its frame.

3. If Constraint Details is not already turned on, turn it on now. This will provide useful details as you are adding your constraints.

4. From the Model tab on the menu bar, select Hinge from the Create button's drop-down menu (Figure 4.11).

FIGURE 4.11
Preparing to select Hinge.

5. Click to place one attachment inside the right door frame and another on the right side of the door where the hinge should be (Figure 4.12).

FIGURE 4.12
Attaching a hinge to the door.

TIP

Line Up the Attachments

Try to line up the two attachments so that the constraint indicator is straight and level. If it's off a little, that's fine, but too much and the door may swing oddly.

For each attachment, the orange rod shows the direction of the hinge, and the orange circle is the hinge's range. Rotate both attachments so the orange rod runs up and down like the one on the left in Figure 4.13 rather than side to side. This will enable the swinging movement you need for the door.

6. Once you have added the hinge, move the door back where it belongs. Scale the door down so there is a tiny gap between the door frame and the door. Otherwise, the door may stick, just like in real life. Playtest to try out your door (Figure 4.14).

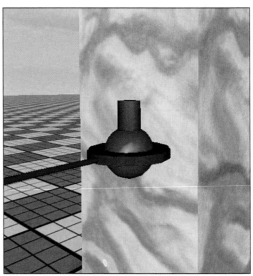

FIGURE 4.13
Rotate the attachments so the orange axis runs up and down, as shown in the image on the left, rather than side to side, as in the image on the right.

FIGURE 4.14
Hinged door opening.

Correctly Rotating the Door

If the attachments are not rotated correctly, your door will swing up and down instead of side to side. If your door is not rotating in the correct direction, you can fix it by rotating both attachments so that the indicator rod is pointing up and down the same way. As you can see in Figure 4.15, the red indicator rod is pointing up and down instead of side to side. Figure 4.16 is an example of correctly oriented hinges, which work side to side.

FIGURE 4.15
Red indicator rod pointing correctly downward.

FIGURE 4.16
Matching indicator rods on both sides of a hinge constraint.

After your door is swinging properly, you need to set the constraint so that the door doesn't spin on its axis. To prevent this, select the HingeConstraint in Explorer, and then go to Properties. Enable LimitsEnabled and then set the LowerAngle to -80 and the UpperAngle to 80 (Figure 4.17).

FIGURE 4.17
HingeConstraint properties with limits set on the UpperAngle and LowerAngle.

Green stoppers appear to indicate where the door should stop depending on the limit's properties (Figure 4.18).

FIGURE 4.18
Green stoppers showing how wide the door will be allowed to swing.

Creating the Springs

A hinge opening adds to the door's realism, and you can complete the effect with an automatic spring closure. SpringConstraint applies a force to its attachments based on spring and damper behavior. This constraint works much like a real-life spring where there is a resting distance, and if the attachments are further than that distance, the spring pulls them together. If they are closer than that distance, the spring pushes them apart.

Follow these steps to add a spring closure to your door:

1. Move the outer part of the left door frame, opposite of the hinge, away temporarily to make it easier to add the springs, as shown in Figure 4.19.

FIGURE 4.19
Door frame opposite of hinge moved away to make constraint creation easier.

2. Click Create and select Spring from the available constraints.

3. Select Spring from the Constraints menu. Click to add a spring attachment from the left side of the door to the exterior door frame (Figure 4.20).

4. Create a second spring attachment by clicking the existing attachment just made on the door and then selecting the exterior door frame on the opposite side. Test your new door when finished. Figure 4.21 shows how the springs should be attached where the springs are the red lines, the hinge is the green dot, and the door frame is the black rectangle.

FIGURE 4.20
Spring constraint attaching the far side of the door to the exterior of the door frame.

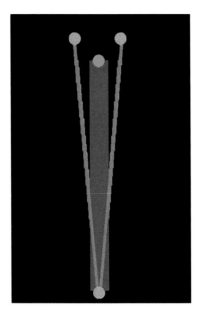

FIGURE 4.21
Diagram of springs (red) and hinge (green) connecting the door to the door frame.

Making the Springs Realistic

To give the springs the illusion of realism, use these steps:

1. In Explorer, select both springs.

2. Make the springs invisible. In Properties, uncheck the Visible option (Figure 4.22). The springs won't be visible during playtesting.

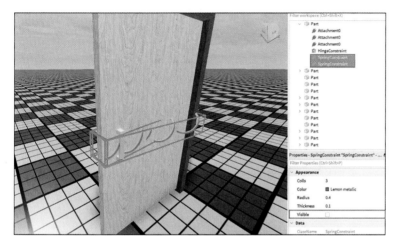

FIGURE 4.22
Both springs selected in Explorer.

3. Make the door realistically spring closed by scrolling down in the Properties window and changing Damping to 850 and Stiffness to 2850 (Figure 4.23).

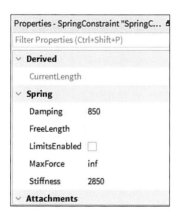

FIGURE 4.23
Damping set to 850 and Stiffness to 2850.

You've made a realistic door that can open on a hinge and automatically spring closed. Playtest and adjust the springs and attachments as needed.

Using a Motor

Let's test what you've learned to build another mechanical construction that you could use in a game: a fan with a working motor. There isn't a separate constraint for motor, but you can use the same HingeConstraint you used earlier in this hour by changing its Properties settings. Follow these steps:

1. Make a fan with parts. Anchor the base, but leave the blades unanchored.

2. Move the fan away from the base in a straight line to make it easy to place the attachments (Figure 4.24).

FIGURE 4.24
Fan blades and base made from parts.

3. Go to Model, click Create, and select HingeConstraint.

4. Place one attachment on the base and the other attachment on the back side of the fan, in the middle, as shown in Figure 4.25. Make sure the attachments are aligned; they need to be straight so they can make a joint and work as a motor.

FIGURE 4.25
Attachments for a hinge constraint in a straight line.

5. Select the HingeConstraint from Explorer, and then from Properties, change the ActuatorType to Motor (Figure 4.26).

FIGURE 4.26
Set ActuatorType to Motor.

6. Also, in Properties, set the AngularVelocity option to 0.6. This is a radians per-second spin speed (rotations per second).

7. To increase the speed of the fan, increase the AngularVelocity. If you want the fan to spin in the other direction, add a minus to the AngularVelocity.

8. Set the MotorMaxTorque to 100,000 (Figure 4.27).

FIGURE 4.27
In Properties, MotorMaxTorque set to 100,000.

9. Move the fan part back in position where it was (Figure 4.28) and playtest to make sure it works.

FIGURE 4.28
The finished fan.

TIP

Troubleshooting the Motor

If the motor doesn't work, you can try again by removing the attachments and hinge and re-adding them. As with the door, the problem may be that you didn't align the attachments correctly, so you should always be mindful of placement.

Summary

In this hour, you created a working door and fan by mastering springs, hinges, and motors, and you learned how to adjust CanCollide to make it possible for a player to collide with or pass through a part.

Q&A

Q. If I place the HingeConstraint on different positions, will it work?

A. Yes, but it may not work as expected. If the attachments are misaligned, the door may swing oddly.

Q. Do constraints work on anchored parts?

A. No, you have to unanchor the part, so the physics engine can detect it and allow the constraints to interact with it.

Workshop

Now that you have finished, review what you've learned. Take a moment to answer the following questions.

Quiz

1. True or False: The anchor parts do not move when you apply constraints on them.

2. True or False: Hinge constraints cannot be used as motors even if you change the settings.

3. True or False: When CanCollide is disabled on an object, parts and players can pass through it.

4. A fan base has to be _____, so the fan doesn't fall.

Answers

1. True. The anchor parts do not move when you apply constraints to them.

2. False. Hinge constraints *can* be used as motors if you change the settings.

3. True. When CanCollide is disabled on an object, parts and players can pass through it.

4. A fan base has to be anchored so the fan doesn't fall.

Exercises

Think of things in the world that people interact with. If you go to a local park, you might find children playing on a see-saw (Figure 4.29). Use what you know about constraints and attachments to make a working see-saw players can interact with.

1. Build a see-saw using parts, as shown in Figure 4.29. Make sure the pillars of the see-saw are anchored, and the see-saw seat is not anchored.

FIGURE 4.29
Example of a see-saw made with parts.

2. Use a hinge constraint to allow the see-saw to move.

TIP

Use Welds to Hold Different Colored Parts Together

Use Weld constraints to hold nonmoving parts together. In the Create menu, select Weld. Then click the two parts to be welded together. Note that if a part is already selected, the first attachment will be created automatically. In Figure 4.30, the light blue wood part on the seat plank is not anchored; it is welded.

FIGURE 4.30
On the right side are two yellow parts and a light blue part held together with weld constraints.

TIP

Only One Hinge Constraint Needed

Only use a hinge on one side of the see-saw because you can run into problems if you have one hinge on each side. Our general recommendation is to build contraptions with the fewest number of constraints possible.

3. **Bonus**: Using what you've learned so far about part creation and physics, create an entire playground for players to enjoy, as shown in Figure 4.31.

FIGURE 4.31
The playground in *MeepCity* by Alexnewtron.

HOUR 5
Building Terrain

What You'll Learn in This Hour:
- ▶ How to use Terrain tools to generate landscape
- ▶ How to use the Edit tab
- ▶ How to use the Region tab

The Terrain tool is useful in creating realistic landscapes with features such as rivers, mountains, and canyons. If you want to re-create a famous national park, for example, you could do so using the Terrain Editor. In this hour, you find out how you can use Roblox's Terrain Editor to create and sculpt beautiful natural landscapes with a variety of materials and how you can use height maps to help speed up that process. Figure 5.1 shows a landscape created with Terrain tools.

FIGURE 5.1
Naturalistic landscape in *Outdoor Ancient Ruins* by Roblox Resources.

Using Terrain Tools to Generate Landscapes

In this hour, we're creating an island to introduce each of the tools and demonstrate how you can use them to generate, edit, and add detail to your environment. By the end of this hour, you will have created an island similar to Figure 5.2 for use in your games. Let's get started.

FIGURE 5.2
Island created in Terrain Editor.

You can find the Terrain Editor in the Home tab of Roblox Studio. When you click Terrain, a new window opens on the left, revealing the individual tools that you can use in terrain design.

You use the Generate tool to create a random landscape out of any combination of available features (called biomes) you choose. You can create watery cliffs, canyons, arctic landscapes, and more. The tool allows you to control the biomes included, where it spawns, and how much terrain to generate. Before you start work on your island, follow these steps to generate a simple landscape:

1. From the New tab, open a baseplate template.

2. Delete the baseplate: In the Explorer window, click the arrow to expand Workspace, select Baseplate, and press the Delete key.

3. On the Home tab, click Terrain to open the Terrain Editor (Figure 5.3); then click Generate.

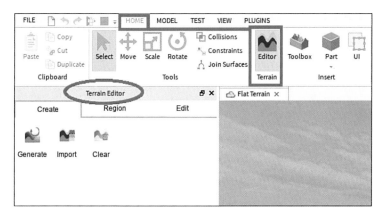

FIGURE 5.3
The Terrain Editor.

The Map Settings section of the Terrain Editor offers a variety of options for the size and position of the terrain to be generated, and the Material Settings section (Figure 5.4) includes the available biomes.

FIGURE 5.4
Generate tool settings.

4. Click the check boxes next to a few biomes you would like to see and click the Generate button. The result for the biomes we chose is shown in Figure 5.5.

FIGURE 5.5
Example landscape built by the Generate tool.

If you decide you don't like the landscape you created, you can click the Clear button (Figure 5.6) to get rid of it and then generate a new terrain.

FIGURE 5.6
The Clear tool.

Begin Work on Your Island

Now that you know about generating a terrain, you can start working on your very own deserted island. Generate a piece of terrain that is only water, as shown in Figure 5.7.

FIGURE 5.7
Generated ocean landscape.

Hint: Make sure any biomes that you don't want aren't selected. For this example, you should select only the Water option. Later in this hour, you find out how to create and adjust the water level.

Using the Edit Tab

With your ocean in place, it's time to add your island and give it some shape. That is where the Edit tab (Figure 5.8), and its tools, come in. These tools enable you to smooth, flatten, erode, and even fill in gaps. For example, if you want to create a cave, you could use the Erode tool to remove terrain. If you want to build a road, you could use the Flatten tool to flatten the terrain before you build on it.

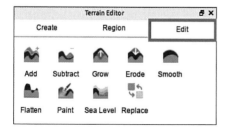

FIGURE 5.8
The Edit tab, including the Add, Subtract, Grow, Erode, Smooth, Flatten, Paint, Sea Level, and Replace tools.

Forming Land with the Add Tool

You use the Add tool to brush terrain into the space. After you click the Add tool, a grid appears in the 3D editor and a blue sphere (the brush) indicates your cursor's position. Click and drag to create terrain wherever you move the brush. The brush is locked to the grid, which is determined by which direction your camera is facing.

Use the Add tool to create the foundation for your island. The one we created is shown in Figure 5.9.

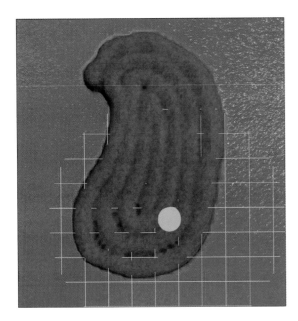

FIGURE 5.9
The Add tool.

You can shape your island however you'd like, but you may find these tips useful:

▶ Adjust the Base Size using the slider to control how much land you are adding.

▶ Click the Top view on the View Selector to look straight down at the water as you draw the shape of your island. The View Selector should be on the top-right of your screen. If you don't see it, go to the View tab at the top of Studio and click the button marked View Selector.

FIGURE 5.10
The View Selector.

Altering Terrain with the Subtract Tool

The Subtract tool is used to remove existing terrain. The Subtract tool works much like the Add tool, except when you click and drag, the tool removes terrain in the shape of the brush wherever the brush currently is (Figure 5.11). The brush is locked to the grid, which is determined by which direction your camera is facing. The Subtract tool's brush settings are the same as the Add tool.

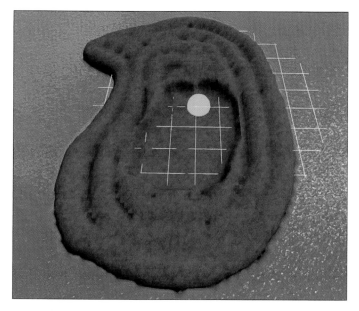

FIGURE 5.11
The Subtract tool.

Use the Subtract tool to create an indentation in the middle of your island for a small lake. You add water to the lake later in this hour.

Elevating Terrain with the Grow Tool

You use the Grow tool to grow terrain on already existing terrain. Try dragging the Grow tool across parts of the island to create mountains and hills (Figure 5.12).

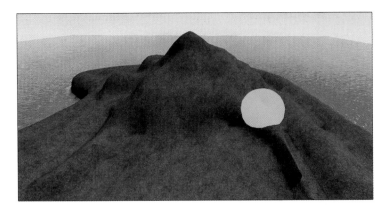

FIGURE 5.12
Use the Grow tool to add terrain such as hills.

The Grow tool's brush settings are the same as the Add and Subtract tools, except the Grow tool also has Strength and Plane Lock (Figure 5.13):

▶ **Strength** allows you to customize how much force the brush will use to grow terrain. The higher the strength, the faster the terrain grows. Turning up the strength allows you to create tall mountains, such as that scene in Figure 5.12, much faster.

▶ **Plane Lock** enables the same grid as the Add or Subtract tool. If Plane Lock is enabled, the terrain only grows along the white grid that appears.

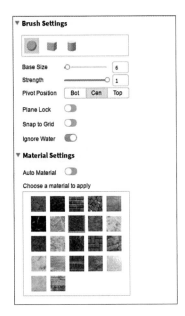

FIGURE 5.13
Grow tool settings.

Removing Terrain with the Erode Tool

The Erode tool is used to remove terrain with an erosion effect. This tool essentially does the opposite of the Grow tool. Clicking and dragging will remove existing terrain in an organic way. Unlike the Subtract tool, this does not remove terrain equally across the entire brush. The Erode tool's brush settings are the same as the Grow tool. Use the Erode tool to create a tunnel to the other side of the hill, as shown in Figure 5.14.

FIGURE 5.14
Use the Erode tool to realistically remove terrain.

Refining Terrain with the Smooth Tool

The Smooth tool is used to smooth messy and spikey terrain. When you click and drag, it will slowly smooth the terrain in its path. The brush settings for the Smooth tool are the same as they are for Grow and Erode. The island is coming along, but you should smooth out some of the bumps created by the Add tool earlier, as shown in Figure 5.15.

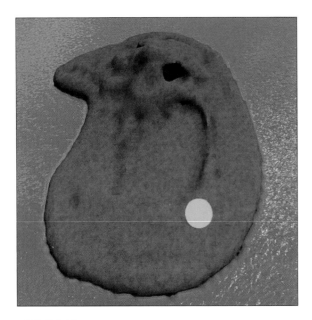

FIGURE 5.15
The Smooth tool evens out erratic terrain.

Flattening Terrain with the Flatten Tool

You use the Flatten tool (Figure 5.16) to make uneven terrain flat, which can be helpful if you want to add cities or roads on top of the terrain. When you click and drag, the tool slowly flattens the terrain in its path. Some areas of the island still need a *bit* more help, so use the Flatten tool to help even out areas.

FIGURE 5.16
The Flatten tool creates a level surface.

The Flatten tool has a couple different settings (Figure 5.17):

▶ **Flatten Mode** allows you to customize whether the tool flattens everything that is above the selection (the left option in the figure), fill everything below the selection (the middle option), or both (the right option).

▶ **Fixed Plane** allows you to set a height within the Plane Position field to base the flattening on. In Figure 5.17, the base height is 30 units high. When disabled, the height is based on the cursor's starting location.

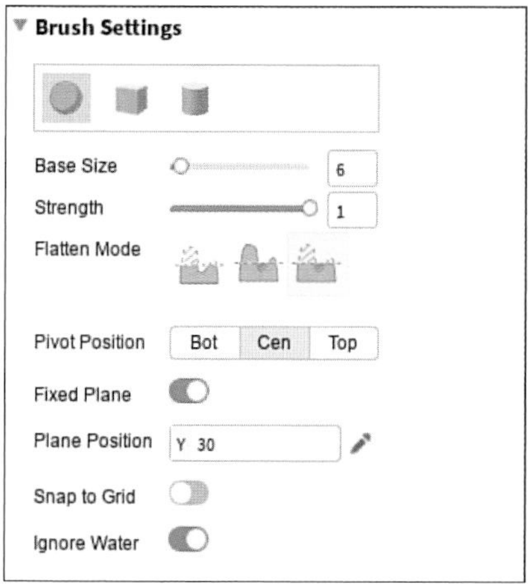

FIGURE 5.17
Flatten tool settings.

Changing Materials with the Paint Tool

You can use the Paint tool to change the material in your terrain. When you click and drag the tool, it replaces the existing material with the one you've selected for the tool. Its brush settings are the same as Grow and Erode. Add some variety to your island by using the Paint tool to turn your shoreline into sand, the bottom of the lake to ground, the entrance of the tunnel to basalt, and the top of the hill into a snowcap, as shown in Figure 5.18.

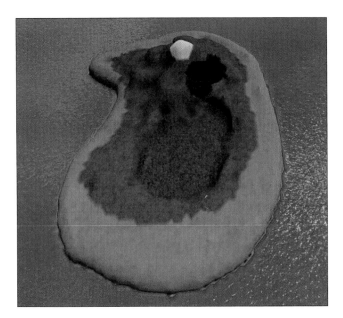

FIGURE 5.18
The Paint tool will change the material of the terrain.

In the Material Settings tab (Figure 5.19), select which material you would like to paint with. Select the sand material and paint along the edges of your island to create beach areas.

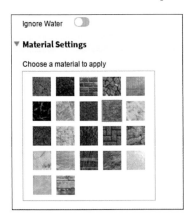

FIGURE 5.19
Menu of available Paint materials.

If you find yourself accidentally changing the water when you don't want to, in Brush Settings, make sure Ignore Water is enabled (Figure 5.20). When this setting is disabled, water can be painted over like any other material.

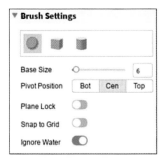

FIGURE 5.20
Ignore Water enabled so that the paint brush is unable to change the material of water.

Creating Water with the Sea Level Tool

Use the Sea Level tool to create a layer of water. To change the sea level, you can either click and drag the blue handles to scale the body of water or manually enter position and size values within the Map Settings tab (Figure 5.21):

▶ **Position** determines the center of where the water should be created.

▶ **Size** allows you to customize how much water it should create in studs.

FIGURE 5.21
The Map Settings tab where size and position for sea level can be input

When you change the settings to your likings, you can click Create to make the water or Evaporate to delete it. Add some water to the space you created for the lake earlier in this hour by increasing the sea level (Figure 5.22).

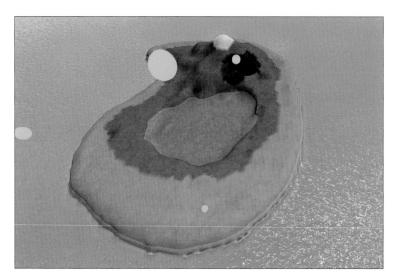

FIGURE 5.22
Adjust the sea level with the Sea Level tool by dragging the blue dots.

Working with the Region Tab

Now that you have the basic terrain for your island, you may want to edit certain sections.
Maybe the mountain feels like it belongs on the south side of the island rather than on the north
side. That is where the Region tab and its tools (Figure 5.23) come into play. Tools in this tab
enable you to work with large landscape areas, which can speed up your creation process.

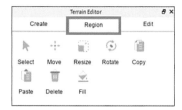

FIGURE 5.23
The Region tab, including the Select, Move, Resize, Rotate, Copy, Paste, Delete, and Fill tools.

Selecting Terrain

You use the Select tool (Figure 5.24) to select terrain. Click and drag to draw a blue box around
everything you want to select (Figure 5.25). If you want to change what you've selected after,
click and hold on the blue handles and then drag them to a new position.

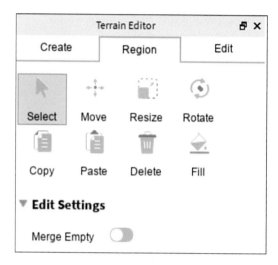

FIGURE 5.24
The Select tool.

FIGURE 5.25
Select terrain with the Select tool.

Moving Terrain with the Move Tool

You can use the Move tool to move selected terrain. When the Move tool (Figure 5.26) is enabled, the blue box from the Select tool becomes white (Figure 5.27). When you click and drag the arrow on any side, the selection moves in that direction. You can also enable Merge Empty under the Move tool (Figures 5.28 and 5.29). Feel free to use the Move tool to change where the mountain is located on the island (Figure 5.30).

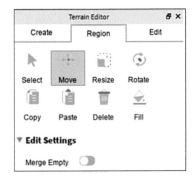

FIGURE 5.26
The Move tool and the Edit Settings tab with the Merge Empty option.

FIGURE 5.27
Original mountain and cave selected within the blue boundaries.

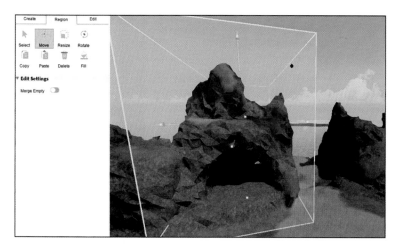

FIGURE 5.28
Disabling Merge Empty maintains the empty space in the cave as the mountain is moved.

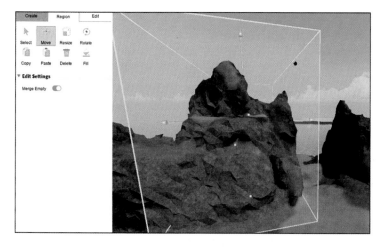

FIGURE 5.29
Enabling Merge Empty allows the empty cave to be filled.

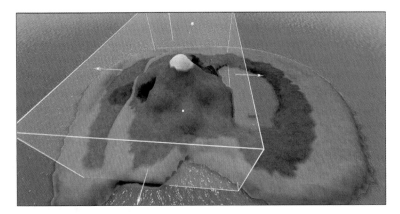

FIGURE 5.30
Move selected terrain with the Move tool.

Scaling Terrain with the Resize Tool

You use the Resize tool to scale the selected terrain (Figure 5.31). When you click and drag the handles on any side of the selection box, the selected terrain resizes in that direction. You can also enable Merge Empty under the Resize tool. Think your mountain is too hill-like? Try increasing its size with the Resize tool.

FIGURE 5.31
Scale terrain with the Resize tool.

Working with the Copy, Paste, and Delete Tools

After selecting an area, you can also choose to create a copy using the Copy tool and paste it in a new location using the Paste tool. You can also delete the selected area using the Delete tool. As you can see in Figure 5.32, the Delete tool removed the mountain and a portion of the lake from the selected area.

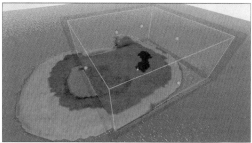

FIGURE 5.32
Before and after use of the Delete tool.

Filling an Area with the Fill Tool

You can use the Fill tool to create a mass of terrain that you can paint, erode, and scale as just discussed, similar to the way a sculptor works with a block of stone (Figure 5.33). It's also handy if you need to create a large block of flat terrain, such as for city blocks.

You can adjust the Material Settings for the Fill tool (Figure 5.34) so that whatever you've selected fills up with a material of your choosing. The Fill tool also has the Merge Empty option.

FIGURE 5.33
A large block created with the Fill tool that can be styled into a mountain using the Erode, Subtract, and Paint tools.

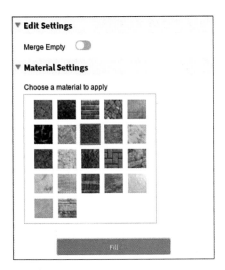

FIGURE 5.34
Material options for the Fill tool.

Using Height Maps and Color Maps

Often, it can be tricky to get exactly the type of terrain that you want and exactly the biomes that you want—that is where height maps and color maps come in handy. These features are particularly useful for re-creating existing natural landmarks or quickly adding specific features. Instead of generating terrain from a specified seed, a height map uses a specified image to determine the height of a given area. Color maps serve a similar function, taking colors from an image to specify a biome for a given area.

Using Height Maps

To manually create a map with specific features, such as a high mountain around a deep valley, can take a lot of time. To speed up this process, you can import a height map. A height map is a 2D representation of a 3D terrain map, as viewed directly from above, as shown in Figure 5.35.

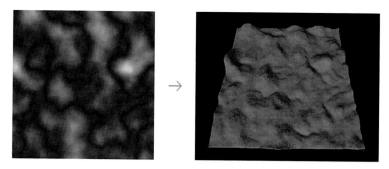

FIGURE 5.35
Example of a height map from Roblox Resources.

A height map allows you to control what every part of your map looks like without having to wait for the terrain to generate every time. When using a height map, it's important to remember that brighter areas of a height map result in higher terrain like mountains, whereas darker areas result in lower regions like valleys.

Use the following steps to import a height map:

1. In the View tab of Roblox Studio, open the Game Explorer or Asset Manager.

2. Click the Import button at the bottom (on the left in Figure 5.36) or the little icon at the top (on the right in the figure) and import your height map picture into Roblox Studio.

FIGURE 5.36
Importing a height map.

3. Go to the Terrain Editor, and under the Create tab, click Import and the text box next to Height Map.

4. Change any properties as you want related to the size and position of the height map, and click Import to have Roblox start generating terrain based off of the height map you imported.

NOTE

Roblox Moderation

The Import button may not be available right away after importing an image. This is because your image needs to go through Roblox moderation first. Be patient, and you should be able to import it after a while (you may need to repeat steps 3 and 4 to refresh). This applies to color maps too, which we cover in the next section.

Using Color Maps

A color map is also a 2D image, but it allows you to specify terrain materials like grass or ice when importing a height map (Figure 5.37). Color maps are useful because you can generate not only the height but also assign materials while it generates. This way, you don't have to worry about having to paint materials on large batches of terrain manually.

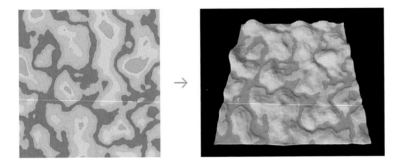

FIGURE 5.37
Example of a color map from Roblox Resources.

Figure 5.38 shows which colors represent which materials.

Color	RGB Value	Material
	[255, 255, 255]	Air
	[115, 123, 107]	Asphalt
	[30, 30, 37]	Basalt
	[138, 86, 62]	Brick
	[132, 123, 90]	Cobblestone
	[127, 102, 63]	Concrete
	[232, 156, 74]	CrackedLava
	[101, 176, 234]	Glacier
	[106, 127, 63]	Grass
	[102, 92, 59]	Ground
	[129, 194, 224]	Ice
	[115, 132, 74]	LeafyGrass
	[206, 173, 148]	Limestone
	[58, 46, 36]	Mud
	[148, 148, 140]	Pavement
	[102, 108, 111]	Rock
	[198, 189, 181]	Salt
	[143, 126, 95]	Sand
	[137, 90, 71]	Sandstone
	[63, 127, 107]	Slate
	[195, 199, 218]	Snow
	[139, 109, 79]	WoodPlanks
	[12, 84, 92]	Water

FIGURE 5.38
The key for materials used in color maps.

Don't worry if the colors on your color map don't match the official color samples exactly. Roblox will try to match the colors as close as possible when it inserts the materials.

Summary

In this hour, you have learned how to use the Terrain Editor and its tools to generate, modify, and sculpt landscapes in Roblox Studio. You also learned about height maps and color maps, which you can use in place of a seed/settings to re-create a map. With these tools, you can create any landscape imaginable—a network of caves, a tropical island, an urban cityscape, an ancient forest, or even Mars; your imagination is the only limitation.

Q&A

Q. Why should I use a height map?

A. A height map enables you to control what every part of your map looks like without having to wait for the terrain to generate every time. You should use a height map if you need a *very* specific feature in your map, such as one tall mountain surrounded by desert or a large swamp in a deep valley.

Q. Why is a color map useful?

A. A color map allows you to assign materials and colors to be used as the height map is generating. It eliminates the need to manually paint materials on large batches of terrain.

Workshop

Now that you have finished, let's review what you've learned. Take a moment to answer the following questions.

Quiz

1. Besides parts and unions, you can use the _____ to build up your game world.

2. True or False: The same seed will produce a map that looks the same each time, regardless of the settings.

3. To generate a map with specific features (such as a high mountain around a deep valley), you can import a _____.

4. True or False: The Add/Grow and Subtract/Erode tools function the same.

5. The _____ _____ tool allows you to control the height at which water generates on your terrain.

6. Materials can be swapped by brushing them with the ____ tool.

7. One way that you can change the materials in your imported map is by also using a _____.

8. True or False: The Add and Subtract tools are locked to a grid that changes depending on your camera angle.

Answers

1. Terrain/Terrain Editor.

2. False. Depending on your settings, you may change how the terrain looks.

3. Height map.

4. False. The Add and Subtract tools apply equally across the brush and don't require existing terrain. Grow/Erode tools do require existing terrain and apply organically to the existing terrain.

5. Sea Level.

6. Paint.

7. Color map.

8. True. The Add and Subtract tools are locked to a grid that changes depending on how you're looking at the surface (the camera angle).

Exercises

This exercise combines a number of different things you've learned in this hour. If you get stuck, don't forget to refer to the previous pages in this hour. Create a map by importing a height map into your game and then adding a twist to it.

1. Find a height map online (be careful to check that it is commercially free to use) and import it.

2. Use the Paint tool in the Edit tab to assign material to the terrain (or use a color map).

3. Continue editing the terrain by using the tools in the Region tab to modify larger pieces of land.

4. Use the tools in the Edit tab to fine-tune your creation.

5. Play around with the sea level. This is especially interesting in landmarks that used to have water (like the Grand Canyon, shown in Figure 5.39).

FIGURE 5.39
Example of modifying an existing landmark such as the Grand Canyon.

It's often easier to plan out a layout for your game outside of Roblox Studio before beginning work on it. Particularly if you're working on a large fantasy-style world, you may want to plan valleys for town areas, mountains to explore, and the path along which players need to travel to get to quests. Using image-editing software such as Photoshop or GIMP, try drawing out the layout for your game world and then importing it using height maps and color maps.

1. Create your own height map and color map with image-editing software like GIMP. They don't need to be fancy—even a simple map can be a good starting place for a world and can save time.

2. Change the sea level to fit your imported terrain. This is useful if you need to add a lot of water in an area, like at the bottom of a lake or river.

3. Modify larger areas of your terrain using the tools in the Region tab.

4. Fine-tune smaller areas using the tools in the Edit tab.

5. Finally, add detail by changing some materials with the Paint tool.

6. **Bonus:** Add grass to your terrain by enabling the Decoration property (Figure 5.40) in your terrain. You can also change the color of materials as you see fit.

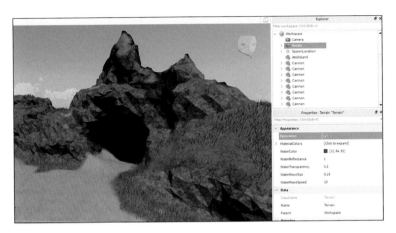

FIGURE 5.40
Terrain with grass enabled.

HOUR 6
Lighting Environment

What You'll Learn in This Hour:

▶ What the properties of world lighting are

▶ How to use lighting effects

▶ How to use SpotLight, PointLight, and SurfaceLight

Now that you have some experience with creating amazing landscapes, it's time to bring light to your worlds. Two important elements for making your world realistic are light and shadows. In this hour, you find out how to use the world-wide Lighting settings to add light—from ambient light to shadows cast from objects—to an environment to make it dynamic and realistic. You use the Lighting settings to make the game brighter, dimmer, or a different color, or you even can add effects such as Bloom, ColorCorrection, Blur, SunRays, and DepthofField. Do you want to create a forest dappled in light or a city glowing with neon signs? These settings enable you to create those worlds. With more effects, you can make the game look more realistic and better match your theme. You also control day, afternoon, and night through lighting. Figure 6.1 shows a scene with areas of sunlight, shadow, and neon light.

With the world-level lighting in place, you can then use additional lighting objects to light up interior spaces or props such as street lamps and flashlights. Let's take a more detailed look at how to add light to games.

FIGURE 6.1
City block featuring realistic world lighting, cast shadows, and neon lights.

Properties of World Lighting

World lighting elevates your game into something dynamic and realistic. This section describes the properties of lighting and how you can use them to bring your games to life. To begin, you need to build a city such as the ones in Figures 6.2 and 6.3. For this tutorial, include a few buildings and a grassy area, but otherwise the creation is entirely up to you.

FIGURE 6.2
Build a city of your own design.

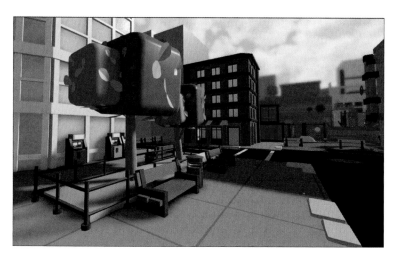

FIGURE 6.3
Cartoony style city example.

With your city built, select Lighting in the Explorer (Figure 6.4). Lighting properties are divided into four categories: Appearance, Data, Behavior, and Exposure (Figure 6.5).

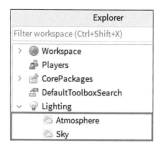

FIGURE 6.4
The Lighting icon in Explorer.

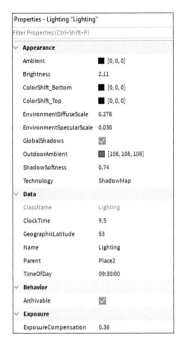

FIGURE 6.5
The properties of Lighting.

Appearance Properties

The Appearance section (Figure 6.6) includes several different properties for customizing your world.

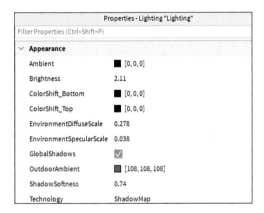

FIGURE 6.6
The Appearance properties of Lighting.

The default settings are a bit dark for a city, so you can adjust them to create a more interesting space:

▶ **Ambient** light is lighting in places that are obstructed from the sky. Ambient light often makes up the base light in a room, such as from ceiling lights or candles. In rooms lit by modern fluorescent lights, the walls take on more of a green tint. When you click the box in the Ambient field, the color-picker opens, and you can adjust the ambiance depending on your game needs.

▶ **Brightness** is the intensity of illumination in a space. To adjust your Brightness, you can enter a value in the field, or you can drag the bar next to the field to make the scene brighter or dimmer.

▶ **GlobalShadows** are enabled by default, so preset shadows are employed to make the environment realistic. You can toggle this feature off if you want your environment to appear starker.

▶ **EnvironmentDiffuseScale** is an ambient light gained from the environment. Making changes to this property affects changes to the Skybox.

▶ **EnvironmentSpecularScale** is used to make certain game parts, like metal, more realistic by giving them a specular shine (Figure 6.7).

FIGURE 6.7
EnvironmentSpecularScale on a building.

To test out these properties, adjust the Brightness of the game. The results should look similar to Figure 6.8.

FIGURE 6.8
A brighter city.

Data and Exposure Properties

In the Data section (Figure 6.9), the two modifiable lighting properties you'll find most useful are the following:

▶ **ClockTime** allows you to change the time of day, and this affects the light and shadows in your environment. Like with the Brightness property, the ClockTime field also includes a bar that you can drag to change the time. Dragging the bar allows you to slowly watch your environment change throughout the hours: You can see the shadows move as the sun rises.

▶ **TimeOfDay** displays the 24-hour clock and automatically adjusts as you manipulate the ClockTime field.

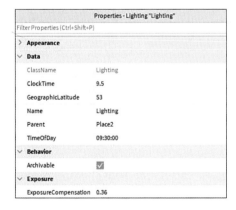

FIGURE 6.9
The Data and Exposure properties of Lighting.

Using Lighting Effects

Lighting effects are used to make the game look more realistic or thematic. For example, the Sun-Rays effect adds rays coming from the sun, which can be decreased or increased by the properties of this effect. This effect can add realism to your sun. Use the following steps to add the SunRays effect to Lighting:

1. From the Explorer, select Lighting and click the plus sign button.

2. From the list that opens, search for the effect you desire. In this case, we're choosing SunRays (Figure 6.10). Click to insert it.

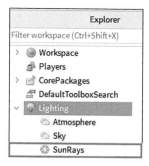

FIGURE 6.10
The SunRays effect added to Lighting.

3. Adjust the Spread and Intensity properties to achieve the desired effect.

NOTE

Rendering Settings

If you are not seeing effects such as SunRays, you may need to turn up the quality of Studio's rendering settings. To do so, go to File, Studio Settings, Rendering and raise both Quality Level and Edit Quality Level.

Figure 6.11 shows a scene with the SunRays effect in use.

FIGURE 6.11
Image without SunRays effect (left) and with SunRays effect (right).

There are several other effects that you can add to make your game look outstanding. To experiment with these effects, build some objects in your city that are made of neon or plastic and include some structures on a grass terrain. The following list describes these effects:

▶ **Bloom effect** increases the glow on lights. It doesn't make the existing light brighter, but it adds more sheen on parts that are plastic and neon. It even adds a glow to the sun and Skybox. Figure 6.12 shows the Bloom effect in action.

FIGURE 6.12
Neon pillars without Bloom effect (left) and with Bloom effect (right).

▶ **ColorCorrection effect** is used to change the environment color. With this effect's properties, you can manage the brightness, contrast, and saturation of the light. This can be used to make your game look more surreal by adding a bit of contrast and changing the tint color. Figure 6.13 shows how you can use the ColorCorrection effect on some structures on grass to enhance the contrast between the object and the natural terrain.

FIGURE 6.13
Image without ColorCorrection effect (left) and with ColorCorrection effect (right).

▶ **BlurEffect** blurs everything the camera sees. This effect can be used to show very hot environments or that the player character is not well. Use the Size property to control how much the world is blurred. Figure 6.14 shows the BlurEffect in action.

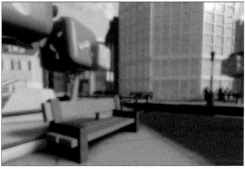

FIGURE 6.14
Image without BlurEffect (left) and with a BlurEffect object with the Size property set to 10 (right).

▶ **DepthOfField effect** is used to bring focus to a specific area and simultaneously blur everywhere else, as shown in Figure 6.15. Adjust the FocusDistance and InFocusRadius properties to leave the desired area in focus and blur the rest.

 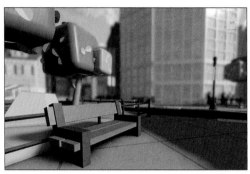

FIGURE 6.15
Image without DepthOfField effect (left) and with DepthOfField effect (right).

Using SpotLight, PointLight, and SurfaceLight

In Roblox Studio, you have three kind of light objects to help you create an even more dynamic game environment: SpotLight, PointLight, and SurfaceLight. You use these light objects to make an area brighter as in real-life scenarios. It's important to make the game bright so players can see what's around them, and these three objects customize the light of objects—the way it is emitted, its range, angle, and more.

Light objects need to be parented to parts (typically simple blocks that are then made transparent) in the game world. Every light object emits light depending on the type of light and its properties. Properties of a light object include the following:

- ▶ **Angle** is where the light is visible.

- ▶ **Brightness** controls how bright the light will be.

- ▶ **Color** controls the color of the light.

- ▶ **Face** sets the side of the light coming from the object.

- ▶ **Range** controls how far the light can reach.

- ▶ **Shadows**, if enabled, reflect shadows if the light is blocked by a part.

All light objects have similar properties, so the preceding options are helpful for setting up every light object.

In our examples, we tested the light objects in an open ground, but in your environment, you can test it in buildings, rooms, or in an open ground as we have.

SpotLight

SpotLight is a cone-shaped light, which is perfect for directional light. You can use it in buildings, car headlights, or flashlights. If you rotate a spotlight, the light rotates with the part. Figure 6.16 shows a nighttime scene lit with a series of SpotLights parented to street light props.

FIGURE 6.16
Image with SpotLight overhead (left) and with SpotLight on the ground (right).

PointLight

PointLight is light that comes from a single point and emanates in all directions instead of a cone. PointLight objects are used to create light sources such as candles and lightbulbs. An example is the lantern lighting in the dungeon shown in Figure 6.17.

FIGURE 6.17
Image with PointLight overhead.

SurfaceLight

SurfaceLight lights up a single face of an object. Examples include the front face of a computer screen or a clock face. Figure 6.18 shows the light from the bottom face of the Roblox Studio logo. Note that unlike SpotLights, the light emanates from the entire face rather than from a single point.

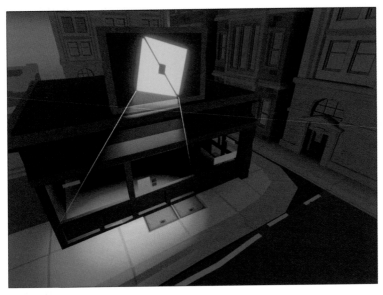

FIGURE 6.18
SurfaceLight on the bottom of the Roblox Studio logo lighting up the area beneath it.

Summary

In this hour, you've learned about world lighting and practiced the lighting settings to make your game look awesome. You learned about lighting objects, which are useful in adding effects to your game. You've seen why it is important to make your game as realistic and dynamic as possible so that it attracts players and is enjoyable to play. You've also seen why light objects are an important game element, not only so that players can clearly navigate a map, but also to add realism to your game objects, such as adding headlights to a car. Try to choose the best settings for your lights so it enhances the details of your world and makes your game something everyone loves to play.

Q&A

Q. Can you add blur with the SunRays effect?

A. No, you cannot add blur with the SunRays effect.

Q. Can you manage contrast from the ColorCorrection effect?

A. Yes, there is a contrast property in the ColorCorrection effect that can be increased or decreased.

Q. Where do you find the Ambient property?

A. The Ambient property is at the top of the Lighting properties.

Q. What does the EnvironmentSpecularScale property do?

A. The EnvironmentSpecularScale property makes the game more realistic by adding a reflection, called a specular, on materials like metal and plastic.

Q. How can you insert the Lighting effect objects?

A. Effect objects can be inserted by clicking the plus sign button on the Lighting and then searching for the effect. Click it to parent the effect to the Lighting.

Workshop

Now that you have finished, let's review what you've learned. Take a moment to answer the following questions.

Quiz

1. The ShadowSoftness property makes the shadow _____.

2. True or False: SpotLight is a directional light.

3. The Bloom effect adds a brightness/glow in the ___.

4. True or False: Blur is even used to blur the background of the GUI.

Answers

1. The ShadowSoftness property makes the shadow "blurry."

2. True. SpotLight is a directional light.

3. The Bloom effect adds a brightness/glow in the "sky."

4. True. Blur is even used to blur the background of the GUI.

Exercises

These exercises combine a number of different things you've learned in this hour. If you get stuck, don't forget to refer to the previous pages in this hour.

In the first exercise, you're creating a perfect spotlight:

1. Find or create a traffic light to act as your light source.

2. Insert a SpotLight object into a part where light would emanate from.

3. Experiment with the Brightness, Angle, and Range properties until you find what works for your scene.

4. Enable Shadows to allow the light to cast shadows.

In the second exercise, try to create lighting for a sunny day:

1. Insert the SunRays effect in Lighting.

2. Change the Intensity property to **0.174** and the Spread property to **0.13**.

3. Insert the ColorCorrection effect and set the Brightness to 0, Contrast to 0.1, and Saturation to 0.

4. Insert the Bloom effect, set the Intensity to 0.5, Size to 53, and Threshold to 1.232.

5. Click the Lighting and add the following properties:

   Ambient to [223, 223, 223]

   Brightness to 6

   ColorShift_Bottom to [255, 255, 255]

   ColorShift_Top to [255, 255, 255]

   EnvironmentDiffuseScale to 0.068

   EnvironmentSpecularSize to 0.748

   GlobalShadows to Enabled

   OutdoorAmbient to [255, 255, 255]

   ShadowSoftness to 1

   Technology to ShadowMap

   ClockTime to −9.727

   GeographicLatitude to −12.732

   TimeOfDay to −9:43:36

   ExposureCompensation to −0.25

For the last exercise, add some sun rays to your scene:

1. From the Explorer, click on "Lighting" and insert the SunRays object.

2. Set the Intensity to 0.375 and the Spread to 0.02.

Your sun should look similar to Figure 6.19.

FIGURE 6.19
An example of the SunRays effect.

HOUR 7
Atmosphere Environment

What You'll Learn in This Hour:

▶ How to use atmosphere properties

▶ How to customize Skybox

You've learned how to make your game environments dynamic through lighting. Now we are going to take this one step further by showing you how to create even more realistic scenes through another effect, Atmosphere. The Atmosphere object uses density and air particle properties to simulate the way sunlight scatters in a real environment (Figure 7.1). Additionally, it controls haze and glare, which is especially useful for creating things such as an exceptional sunrise, morning fog, deep space, and a lot more. In this hour, you find out how to use atmosphere properties and create a custom Skybox, which provides additional atmosphere in your game environments.

FIGURE 7.1
The atmospheric setting of template *Galactic Speedway*.

Using Atmosphere Properties

To use the atmospheric settings, you need both a Sky and an Atmosphere object parented to lighting. To add Sky and Atmosphere objects in one of your existing places, do the following:

1. Select Lighting in Explorer. If Sky and Atmosphere are not already available under Lighting, as shown in Figure 7.2, use the plus button to insert these objects.

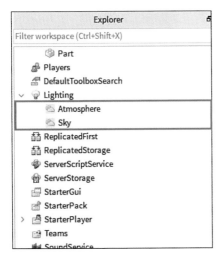

FIGURE 7.2
The Sky and Atmosphere objects in Explorer.

2. Click the Atmosphere object. The properties of Atmosphere appear in the Properties window (Figure 7.3). The following sections talk about these properties in detail.

TIP

Setting Properties to the Defaults

With all of the properties of Atmosphere, check whether they already exist in your game and have been modified. If so, you may need to delete them and insert them again so the settings are back to defaults. Then you can successfully achieve the following atmospheres.

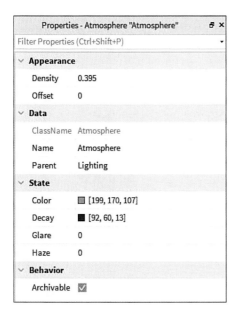

FIGURE 7.3
Properties for the Atmosphere object.

Density

Density defines the number of particles in the atmosphere. In a very dense environment, such as a tranquil forest, objects or terrain in your game will be obscured by air particles, whereas in a bright desert setting where the air is thin, the objects and terrain will be clearly visible.

NOTE

Density and the Skybox

Density does not directly affect the Skybox. It merely affects in-game objects and terrain, leaving the Skybox unchanged and, therefore, still visible.

Adjust the Density property in your scene and compare the difference before and after the adjustment. In Figure 7.4, the Density property is zero, so the image is perfectly clear. In Figure 7.5, the Density property is set to 0.395, and the air is thicker.

FIGURE 7.4
View of Density = 0.

FIGURE 7.5
View of Density = 0.395.

Play around with the density of your environment until your atmosphere has the look and feel that you want.

Offset

The other Appearance property for atmosphere is Offset, which controls how light transmits between the camera and sky. In Figure 7.6, where the offset is zero, notice how the horizon is barely visible, and distant objects are blended into the sky. This gives the illusion of a seemingly endless world. The effect is more noticeable the farther into the distance the horizon is.

FIGURE 7.6
View of Offset = 0.

In Figure 7.7, we've increased the value of offset to 1, which has enforced the silhouette of the horizon against the sky.

FIGURE 7.7
View of Offset = 1.

Offset should be balanced against density and carefully tested in your environment. A low offset may cause "ghosting," where the Skybox is visible through objects or terrain. You can correct this by increasing the offset, which more clearly silhouettes distant objects/terrain against the sky. However, too much offset may reveal level-of-detail "popping" for distant terrain and meshes.

Haze

Haze reduces the clarity of the particles in the sky. In the real world, haziness is generally caused from particles such as dust or smoke. In Figure 7.8, the scene has a low haze value. Remember, density is the number of particles; haze is the clarity of those particles.

FIGURE 7.8
View of Haze = 1.

If we modify the level of haze, it produces a visible effect both above the horizon and into the distance. This can be combined with a change to the Color property to create an environmental mood. For example, if you want to build a dystopian and polluted city, you can modify haze and color to create a smoky tint, such as in Figure 7.9.

FIGURE 7.9
View of Haze = 2.8.

Color

Color changes the atmosphere hue for a subtle environmental mood. As described earlier, color is most outstanding when combined with increased haze to expand the visible effect. In Figure 7.10, the bright blue color indicates a cheerful summer day, and in Figure 7.11, the darker hue gives a more somber effect.

FIGURE 7.10
View of Color = [255, 255, 255].

FIGURE 7.11
View of Color = [250, 200, 255].

Glare

Glare is the atmospheric glow around the sun. Note that the position of the sun is controlled by the time of the day that you've set in Lighting properties. In Figure 7.12, the glare is set to zero, and you can see in Figure 7.13 that increasing the value of glare results in more sunlight cast onto the sky and world.

FIGURE 7.12
View of Glare = 0.

FIGURE 7.13
View of Glare = 1.

TIP

Glare Needs Haze

Glare must be combined with a higher Haze level than zero to see any changes. Without Haze, Glare will not work.

Decay

Decay defines the hue of the atmosphere away from the sun. This effect moves across the sky based on ClockTime or TimeOfDay (which we discuss in more detail later in this hour). In Figure 7.14, the decay is set to white (RGB value 255, 255, 255), and when you modify that value in Figure 7.15, you notice the change in the hue of the atmosphere.

FIGURE 7.14
View of Decay = [255, 255, 255].

FIGURE 7.15
View of Decay = [255, 90, 80].

NOTE

Haze and Glare to Have Decay

Decay must be combined with Haze and Glare to see changes. The level of Haze and Glare must be higher than zero to see any effect; otherwise, it will not work.

Customizing Skybox

Skyboxes can add atmosphere to your game environment or even give the impression that your game world is in deep space or underwater (Figure 7.16). Skyboxes are used to match the theme of your game. You can use a Skybox from Toolbox for free by searching Skybox, or you can make your own Skybox with the instructions in the following sections.

FIGURE 7.16
Skybox and celestial bodies in template *Move It Simulator*.

Making a Skybox

Skyboxes are made up of six individual images, which are wrapped into a cube. A convincing Skybox appears to be panoramic because the images are made and sized to be perfectly aligned with each other. This lets you look all around without the impression of being inside of a cube. Figure 7.17 shows how the six images work together to make a panoramic image.

Making Skybox images from scratch goes beyond the scope of this hour. You must create the images yourself while keeping in mind that each image must be seamless along all edges of the neighboring images so they work together when "folded" into a cube.

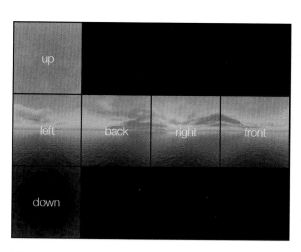
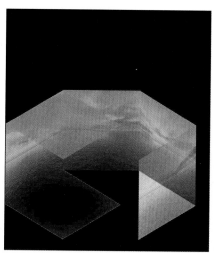

FIGURE 7.17
Six images (left) piece together to make a panorama (right).

Once you have made the Skybox images, do the following:

1. Access Lighting from Explorer.

2. Click the plus button; then click on the Sky object (Figure 7.18).

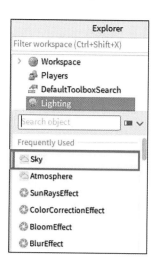

FIGURE 7.18
Parenting the Sky object to Lighting.

3. Once you have parented Sky with Lighting, click the Sky object, and the properties appear in the Properties window. Figure 7.19 shows the names of the six properties as they appear in the Properties box, and Figure 7.20 shows the arrangement of the images.

FIGURE 7.19
The six properties.

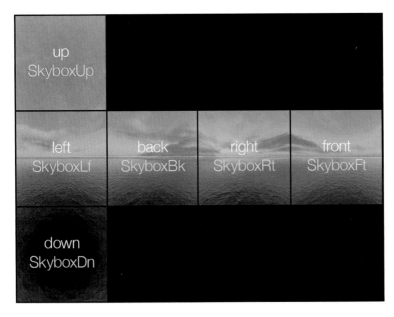

FIGURE 7.20
The arrangement of the images.

4. Click the field next to each of the six Skybox image properties. Select an already-uploaded image or click Add Image. If you're not given the option to upload an image, publish your game and try again.

5. From the pop-up context box, click the Choose File button (Figure 7.21), and select the image for the Skybox. Keep in mind the Skybox-Template and make sure you're uploading the correct image in the correct property.

6. After uploading, click Create.

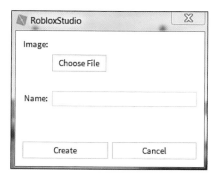

FIGURE 7.21
Click Choose File to upload an image for use.

If you did everything correctly, a complete Skybox will appear in your environment.

Customizing Celestial Bodies

By default, the Roblox sky contains celestial bodies like a sun, moon, and stars. These bodies dynamically rise and set based on the TimeOfDay and ClockTime values, which are in the Lighting properties.

Celestial bodies can be customized as the following:

▶ **Sun:** You can change the image by uploading an image of a new sun to the SunTextureId property (Figure 7.22). You can adjust its relative size with the SunAngularSize property.

FIGURE 7.22
Uploading a new sun image to Explorer.

▶ **Moon:** You can change the image by uploading a new moon image to the MoonTextureId property, and you adjust its relative size with the MoonAngularSize property.

▶ **Star:** You can't change this image, but you can increase or decrease the number of stars with the StarCount property.

Celestial Bodies sets whether the sun, moon, and stars will show or not. To disable all celestial bodies, you can turn off the CelestialBodiesShown property (Figure 7.23). Alternatively, you can disable the sun and moon by setting SunAngularSize or MoonAngularSize to 0.

FIGURE 7.23
Toggle celestial bodies on and off in the Appearance section.

Adjusting Lighting Color

In real life, the ambient color of light changes throughout the day. For example, sunlight in the early morning or late afternoon is normally warmer with more pink-orange in its tone.

You can access the Ambient property by going to Explorer, Lighting; the Ambient property is at the top of the list.

You can change the Outdoor Ambient color by selecting the Color property and choosing which color you want, but making the change here doesn't change the entire theme color of the game. To make that change, you have to change the ColorShift_Top property as well; you find that under the Ambient property. Figures 7.24 through 7.27 show some examples.

FIGURE 7.24
View of the Sunrise theme (Ambient = [255, 100, 150], ColorShift_Top = [255, 100, 150]).

FIGURE 7.25
View of the Sunset theme (Ambient = [255, 100, 0], ColorShift_Top = [255, 100, 0]).

FIGURE 7.26
View of the Cloudy Sky theme (Ambient = [110, 110, 130], ColorShift_Top = [110, 110, 130]).

FIGURE 7.27
View of another Cloudy Sky theme (Ambient = [110, 110, 225], ColorShift_Top = [0, 150, 225]).

Summary

In this hour, you've learned about atmosphere's properties—Density, Offset, Haze, Color, Glare, and Decay—which are used to make the environment more enjoyable by adding realistic detail. You have learned about Lighting, Ambient, and ColorShift_Top and how you can use these properties for scenes such as sunsets, underwater worlds, blue night, and many other light tones. Lighting is one of the main components that make your game look outstanding. Great lighting reveals dynamic details of your work and increases the definition of your game.

Q&A

Q. Can the Ambient property change the whole game light tone?

A. No, you have to change the ColorShift_Top color as well to change the whole game light tone.

Q. Does a high level of Glare affect the sun?

A. Yes, it makes the sun appear larger and increases the sunlight.

Workshop

Now that you have finished, let's review what you've learned. Take a moment to answer the following questions.

Quiz

1. True or False: Glare must be combined with a Haze level higher than 0 to see any effect. Without Haze, Glare may not work.

2. True or False: Decay defines the haziness of the atmosphere with a visible effect both above the horizon and into distance.

3. Skyboxes are made up of __ individual images which are wrapped into a cube.

4. True or False: Skyboxes can be replaced.

Answers

1. True. Glare must be combined with a Haze level higher than 0 to see any effect. Without Haze, Glare may not work.

2. False. Haze defines the haziness of the atmosphere with a visible effect both above the horizon and into distance.

3. Skyboxes are made up of six individual images which are wrapped into a cube.

4. True. Skyboxes can be replaced.

Exercises

See if you can create a gritty dystopian sky (Figure 7.28) using the city scene you created previously or one of the Roblox templates. Ask yourself some questions: Is your city on earth? Or is it on an alien planet? Change up the celestial bodies and the color of the atmosphere to match.

FIGURE 7.28
Modified version of *Beat the Scammers*.

1. Using the city scene you created in Hour 6, upload an image of a new sun to SunTextureId.

2. Modify the ClockTime property to make the time of day at sunset.

3. Play with the Haze, Color, Decay, and Glare properties to make your cityscape appear dystopian.

HOUR 8
Effects Environment

What You'll Learn in This Hour:

▶ How to use particles
▶ How to use beams

Particles are used to create effects, such as a trail of stardust, orbs of light falling through a forest, and even leaves blowing in the wind. This effect helps games feel more alive and immersive for players. For example, you can create a fireplace, add a fire particle, and perhaps even add some smoke and sparkles, and suddenly your player is cozied up to a relaxing fire. You can further customize your own particles by adding a texture.

In addition to using particles to create an effect, you can use a Beam object to connect two attachments. Beams have a constant particle going from one to the other, and you can use them on light objects to make them look realistic.

In this hour, you will learn how to customize your Roblox games using these particles and beams. Read on to find out how these effects make your game more enjoyable to play, thereby increasing engagement with users.

Using Particles

Particles are unique effects that attract players and have unlimited uses. You can use them to create smoke, fire, sparkles, rain, waterfalls, and custom particles. Figure 8.1 shows an environment that uses particles.

FIGURE 8.1
Particles used to create a volcanic effect in template *Pirate Island*.

To practice using particles, do the following:

1. Make a part.

2. Click the plus button and insert the ParticleEmitter object (Figure 8.2).

FIGURE 8.2
Inserting a ParticleEmitter object.

Rate controls how fast particles spawn across the part's face. If the parent part moves, the particles create a trail by default. If you make the part bigger, particles spawn over a larger area, but the rate of the particles stays the same. If you make the part smaller, the number of particles is more closely packed together. (See Figure 8.3.)

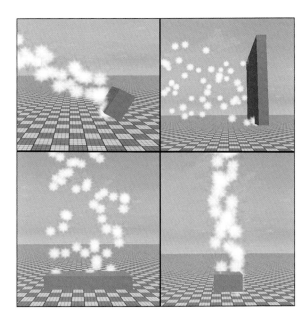

FIGURE 8.3
The same number of particles as emitted from parts of different sizes.

If you don't want to move the part, you can use EmissionDirection property to change the direction of the particles. You access that property by clicking ParticleEmitter object and then changing the direction in the properties.

Customizing Particles

You can easily customize particles by adding a texture to the ParticleEmitter object. To accomplish this, do the following:

1. Add a ParticleEmitter to the part.

2. Click the object, and the properties of the ParticleEmitter appear in the Properties window.

3. Make sure your game is published, then click the Texture property and add a texture (Figure 8.4). *Note the background of the texture must be transparent.* The texture of the particles changes, as shown in Figure 8.5.

Properties - ParticleEmitter "ParticleEmitter" 5	
Filter Properties (Ctrl+Shift+P)	

⌄ **Appearance**	
Color	[255, 255, 255]
LightEmission	0
LightInfluence	1
Size	1
Texture	...t_PNG_Clipart_Image-781
Transparency	0
ZOffset	0

FIGURE 8.4
Customize particles by adding a texture.

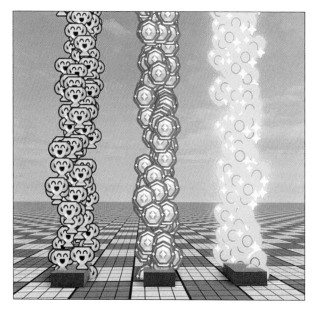

FIGURE 8.5
Particles with added textures.

Changing the Color of Particles

You can tint particles with any color you want. To change the color, do the following:

1. Click the ParticleEmitter, and the properties appear.

2. Click the Color property (Figure 8.6).

FIGURE 8.6
Changing the color of a particle.

3. Choose any color you want from the color picker, and it will overlay the existing color.

4. Click OK to change the color.

Properties of a ParticleEmitter

Like any object, the ParticleEmitter has various customizable properties. Here, we define a few of the most useful:

▶ **Color:** Allows you to add color to the particles.

▶ **LightEmission:** Adds brightness to the particles.

▶ **Size:** Controls the size of the texture. If you increase this, it will make the particle bigger.

▶ **Drag:** How fast particles lose speed.

▶ **Lifetime:** Determines how long the particles last before disappearing.

▶ **Rotation:** Rotates the texture.

▶ **RotSpeed:** Spins the texture coming out of the particles. If you increase this property's setting, the texture spins clockwise. If you decrease the setting, the texture spins counterclockwise.

▶ **SpreadAngle:** Spreads the particles in the direction you want.

Using Beams

A Beam object is a ribbon of texture that can be animated or still to make realistic effects. You can use beams to make a laser, waterfall (Figure 8.7), or even a path.

FIGURE 8.7
Beams used to create a waterfall effect in template *Galactic Speedway.*

To use a beam, you just need to place attachments between two parts, add a texture, and then set speed, transparency, and width. Try it out:

1. Create two parts and keep some distance between them, as shown in Figure 8.8.

FIGURE 8.8
Create a Beam object, starting with two parts.

2. Select one of the parts, click the plus button, and add a Beam object.

3. For both parts, click the plus button and insert an attachment (Figure 8.9).

FIGURE 8.9
Use the Insert menu to add an attachment to each part.

4. Select the Beam object, and in Properties, select Attachment0 (Figure 8.10). You'll notice your cursor changes.

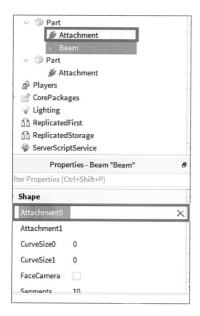

FIGURE 8.10
With Beam selected, in Properties, scroll down and click Attachment0.

5. Click one of the attachments you just made. This will be the starting point of the beam.

6. To set the end point, in Properties, select Attachment1 and then select the second attachment.

7. Once you're done setting the starting and ending points, scroll up and click the Texture property and add a texture to the object.

When you're done, you see the selected texture running between your two parts. In Figure 8.11, a striped image was used as a texture.

FIGURE 8.11
Two parts with a beam between them.

If the parts connected to the beam are moved, the beam stretches and moves to follow.

The Curve Property

The CurveSize properties control how curved the beam strip may be. The higher the number of this property, the more pronounced the curve is on that side of the beam. In Figure 8.12, Curve-Size0 set is set to 10.

FIGURE 8.12
View of the CurveSize0: 10.

In Figure 8.13, the curve is modified from both parts.

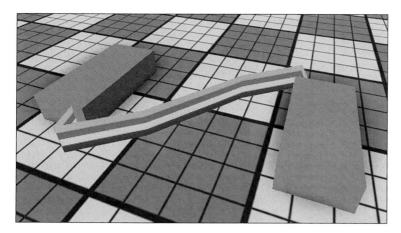

FIGURE 8.13
View of CurveSize0: 10 and CurveSize1: 10.

The Segments Property

The Segments property determines how smooth the curve should be. Increased segments results in a smooth curve, as shown in Figure 8.14.

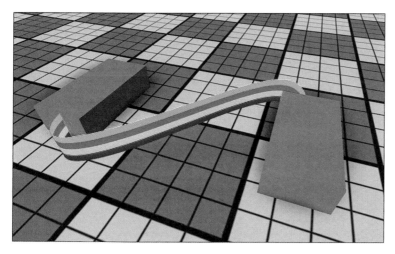

FIGURE 8.14
View of Segments: 60.

Decreased segments results in a rough curve, as shown in Figure 8.15.

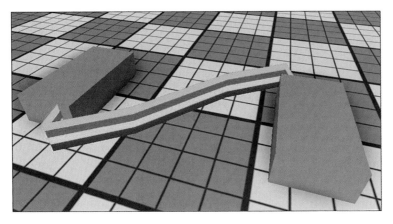

FIGURE 8.15
View of Segments: 5.

The Width Property

Width properties control the size of the beam strip. There are two ends and two properties, and each controls the size of its own end. Keep the number of the properties equal if you don't want one end to be smaller. Figures 8.16 and 8.17 show some image examples of different widths.

FIGURE 8.16
View of Width0: 10 and Width1: 5.

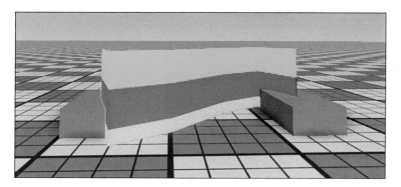

FIGURE 8.17
View of Width0: 5 and Width1: 5.

Adding a Ray Effect on Light with Beam

Say you want to add a ray effect with a beam like Figure 8.18. Spotlights do exist, but stylistically beams can provide more punch. To achieve the effect of Figure 8.18, you need to create a beam that is wider on one end and doesn't have animation. To accomplish this, first add a light ray effect texture to the Beam object and make sure it's transparent from the background. Then apply the following settings:

▶ LightEmission: 0

▶ LightInfluence: 0

▶ TextureLength: 19

▶ TextureMode: Wrap

▶ TextureSpeed: 0

▶ Transparency: 0.5

▶ ZOffset: 0

▶ CurveSize0: 0

▶ CurveSize1: 0

▶ Face Camera: Disabled

▶ Segments: 100

▶ Width0: 3

▶ Width1: 22

With these settings, you will successfully get a light ray.

FIGURE 8.18
Ray effect on light with a beam.

Summary

In this hour, you've learned about particles, which can be used in many places, including creating fire in a fireplace, smoke in a chimney, and sparkles in a box full of treasure. If you add ParticleEmitters to your game, they enhance your game by making it much more realistic for players. You even learned that you can customize particles and change their colors. You also were introduced to using a beam, which is a pretty special effect that renders a texture between two attachments. By using effects like particles and beams, you can create a more immersive environment that will lead to a player's increased engagement in your game.

Q&A

Q. Can you curve the strip of a beam?

A. Yes, you can curve the beam with the curve properties.

Q. Is there a limit to how much you can spread particles over an area?

A. The area over which the particles spawn is controlled by the size of the part. To increase the number of particles spawning, you need to increase the rate property. If you max out the rate and still want more, you can place multiple emitters in the same area. However, this may affect performance.

Q. Can you change the color of the particles?

A. Yes, you can change the color of the particles with the top Color property. This works best on particles with lighter-colored textures.

Workshop

Now that you have finished, review what you've learned. Take a moment to answer the following questions.

Quiz

1. True or False: You can change the color of the beam.

2. True or False: Particles float up from the part's emission direction by default.

3. True or False: Particles are inserted the same way as beams.

4. Beam _____ speed can be controlled from the properties.

5. Particles' spawn intensity can be increased and decreased by the _____ property.

Answers

1. True. You can change the color of the beam.

2. True. Particles float up from the part's emission direction by default.

3. True. Particles are inserted the same way as beams.

4. Beam texture speed can be controlled from the properties.

5. Particles' spawn intensity can be increased and decreased by the Rate property.

Exercises

This exercise combines a number of things you learned in this hour. If you get stuck, don't forget to refer to the previous pages for information.

1. Open Roblox Studio.

2. Build a fireplace such as the one shown in Figure 8.19.

FIGURE 8.19
A fireplace.

   3. Use a combination of fire and smoke particle emitters to make the fire come alive, as in Figure 8.20.

FIGURE 8.20
A roaring fire in the fireplace.

In this second exercise, you make a waterfall. To create a Beam effect, do the following:

1. Place two parts so that they can act as the top and the bottom of the waterfall.

2. Add a waterfall texture to the beam; you can easily find one from Google. The background of the texture has to be transparent.

3. Experiment with the beam's properties, such as curve, width, and segments, to shape the beam. Your finished product should be similar to Figure 8.21.

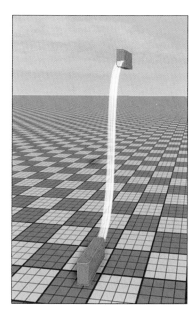

FIGURE 8.21
A waterfall.

TIP

Enhancing the Waterfall Effect

To add more effect to the waterfall, you can add smoke to the part, which will look like mist at the bottom of the waterfall (Figure 8.22).

FIGURE 8.22
Adding mist to the waterfall.

HOUR 9
Importing Assets

What You'll Learn in This Hour:

▶ How to insert and upload free models
▶ How to import MeshParts
▶ How to import textures
▶ How to upload sounds to Roblox and import them in Roblox Studio

All of the game assets—including models, scripts, textures, and audio files—are saved online to Roblox. Unlike many other engines, there is no local game asset storage for either the player or developer. This benefit enables better team collaboration and alleviates storage concerns for players with older devices.

All assets have an individual ID connected with a Roblox account. Once uploaded, an asset is automatically submitted to Roblox's Moderation Team. Moderation usually takes a few minutes, and then your asset will be approved and will appear in the Roblox Studio.

Inserting and Uploading Free Models

Free models are a group of objects that can be treated as a single item in Roblox. For instance, a weapon model might include a group of visible parts, a particle-emitter, attachments, and a script allowing it to function. These are made by Roblox users and they're uploaded to the Asset Library.

Models can be used for sharing assets with friends or making them free for everyone by clicking the Allow Copying option. Once a model is created and uploaded to Toolbox, the version cannot be updated or deleted from your inventory. However, the title, ownership, and description can be updated. The public models that you have in your inventory can be removed.

To create a model, do the following:

1. Build your creation out of parts, and when it's complete, select all parts.

2. In the Home tab, group the parts in the asset by clicking the Group button (Figure 9.1).

FIGURE 9.1
The Group button.

3. Your model is now displayed in Explorer, and you can rename it.

4. Designate the part of the model that will be used as a base when someone is positioning the model by selecting PrimaryPart in the model's Properties window. Once the cursor is active, choose which part in the model you want to be the PrimaryPart (Figure 9.2) and click it to designate it.

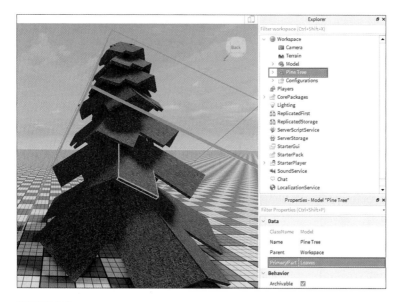

FIGURE 9.2
Select the PrimaryPart.

Uploading the Model to Roblox

Use the following steps to upload a model to Roblox:

1. Once you have a model with a PrimaryPart selected, right-click the model in Explorer and select Save to Roblox from the context menu (Figure 9.3).

⟳	Cut	Ctrl+X
▤	Copy	Ctrl+C
	Paste Into	Ctrl+Shift+V
▤	Duplicate	Ctrl+D
✕	Delete	Del
	Rename	F2
▣	Group	Ctrl+G
▥	Ungroup	Ctrl+U
▥	Select Children	
▣	Zoom to	F
▣	Select Connections	Alt+C
▣	Select Connection 0	Alt+1
▣	Select Connection 1	Alt+2
▣	Swap Attachments	
◎	Insert Part	
	Insert Object...	
	Insert from File...	
	Convert to Package...	
	Save to File...	
	Save to Roblox...	
	Save as Local Plugin...	
	Publish as Plugin...	
	Export Selection...	
⑦	Help	

FIGURE 9.3
Save to Roblox.

2. A context window opens (Figure 9.4). Complete the necessary details:

▸ **Creator drop-down menu:** Me means the model will be saved in your inventory. If you have permissions on a group, you also have the option to save the model to Group Models.

▸ **Allow Copying switch:** By default, the switch is disabled, which means no one else can use the model. Only you can use it. If you enable it, the button turns green, and the model becomes a free model, which means everyone in Roblox can use it.

3. Once you are done with permissions, click the Submit button, and the model starts loading. You should see a confirmation screen that the model has successfully saved to Roblox (Figure 9.5).

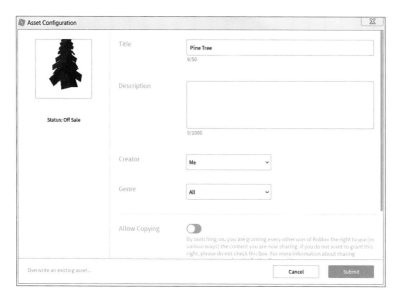

FIGURE 9.4
Complete the description.

FIGURE 9.5
Successfully submitted.

Accessing Models

To access a model, open Toolbox and select the My Models option from the drop-down menu. The asset you saved in Roblox will be there (Figure 9.6).

FIGURE 9.6
My Models in Toolbox.

Inserting Free Models

Free models are models created by the community for other users to freely grab and use in their games.

NOTE

Scripts in Free Models

When using free models, you may want to remove all associated scripts. The code in scripts may be incompatible with your game, or the code may simply be inconvenient. For example, one free model spawns birds every few minutes, and another sets surrounding objects on fire.

To insert a free model in your game, use the following steps:

1. Open Toolbox (Figure 9.7).

FIGURE 9.7
Models in the Toolbox.

2. Use the drop-down menu in the top left of the Toolbox to select the Models category of assets (Figure 9.8). Here can find assets like cars, trees, or anything else you want for the game.

FIGURE 9.8
Models.

3. Click the model you want to insert (Figure 9.9), and the object appears in your Studio. If you want details on the model, click the magnifying icon and the asset opens (Figure 9.10).

FIGURE 9.9
The asset.

FIGURE 9.10
Asset details.

Audio Assets

You can do the same with Audio options. Click the magnifying icon for an audio asset to listen to the audio. Read more about importing audio assets later in this hour.

Importing with MeshParts and Asset Manager

A MeshPart is a physically simulated mesh that supports the upload of meshes in FBX or OBJ format. The simplest way to import meshes to your game is via the Asset Manager window. To import a mesh by MeshPart, do the following:

1. In Explorer, hover over Workspace and click the plus button (Figure 9.11).

FIGURE 9.11
Triggering the plus button next to Workspace.

2. Scroll to the Parts section and select MeshPart (Figure 9.12). The MeshPart appears at the center of your camera view (Figure 9.13).

FIGURE 9.12
MeshPart.

FIGURE 9.13
MeshPart in the center of camera view.

3. Select the MeshPart. To import a mesh, go to Properties, click the folder icon next to MeshId (Figure 9.14), and select a mesh to upload.

FIGURE 9.14
The MeshId.

NOTE

Mesh Size

Meshes must be under 10,000 triangles.

Once you have attached a mesh, the mesh starts loading to the Studio.

NOTE

Moderation for Assets

All the assets uploaded to Roblox go through moderation. Occasionally this results in a slight delay before the asset is available in Studio. If the object does not upload, try making sure there are no numbers in the filename and look for words in the filename that could be triggering moderation.

Bulk Importing Meshes by Asset Manager

The Asset Manager window enables you to manage and access meshes, images, places, and packages in your game. It also enables you to bulk import groups of game assets at once. Use the following steps to bulk import meshes:

1. In the View tab, click Asset Manager (Figure 9.15).

FIGURE 9.15
Asset Manager.

2. Click the Import button (Figure 9.16) to import meshes.

FIGURE 9.16
The Import button on the far right.

3. Once you have uploaded the meshes, a context menu opens. Click the Apply All button (Figure 9.17).

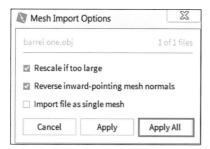

FIGURE 9.17
The Apply All button.

4. The Bulk Import window opens (Figure 9.18) so you can track the progress of the import. Once all the meshes have a green check mark, you can close the window.

FIGURE 9.18
The Bulk Import window.

5. From the Asset Manager, click the Meshes folder. Right-click the mesh and then select Insert with Location from the context menu (Figure 9.19). Then the mesh is successfully added to your game.

FIGURE 9.19
The Insert with Location command.

Importing Textures

There are two ways to import images to Studio. The first way is via Asset Manager, and the second way is by using Texture Object from Explorer. You can import images to Roblox in PNG, JPG, TGA, and BMP format. Via Asset Manager, you can import a large number of textures. To import a texture, do the following:

1. From Explorer, click the plus button (Figure 9.20) and select Texture (Figure 9.21).

FIGURE 9.20
The plus button.

2. Place the texture where you want (Figure 9.22).

FIGURE 9.21
Selecting Texture.

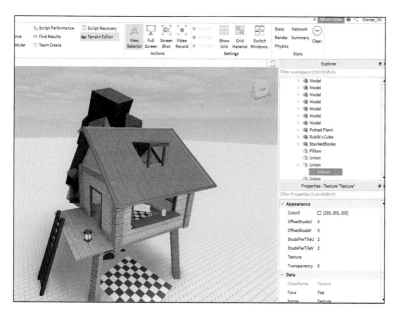

FIGURE 9.22
Placement of the texture.

3. Go to the Properties window, click Texture (Figure 9.23) to open the Image selection dialog box, and then click Add Image.

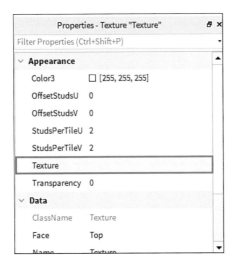

FIGURE 9.23
The Texture option.

4. Use the context window to attach a file and then click Create (Figure 9.24). The texture will be successfully applied.

FIGURE 9.24
Selecting a file.

Importing Decals by Asset Manager

You can import decals in Asset Manager by opening Asset Manager, clicking the Import button, and uploading images. Once you've loaded the image, go to the Images folder, double-click on the image, and place it wherever you want. Figure 9.25 shows decals in Asset Manager.

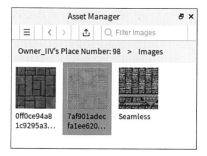

FIGURE 9.25
The Asset Manager with decals.

Importing Sounds

You can import audio files in MP3 or OGG format. Note that you have to pay Robux to upload audio files, but sound can add a lot of entertainment value to your game. The amount depends on the length of the audio, and the fee goes toward the cost of moderating these assets. To upload your own audio file, do the following:

1. Go to the Create page on the Roblox website. On the left side, click Audio (Figure 9.26).

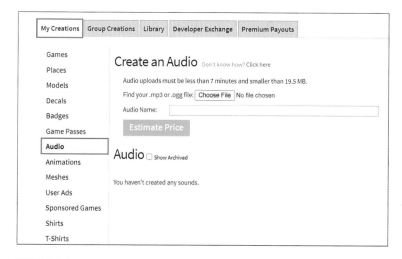

FIGURE 9.26
The Audio option.

2. Attach a file. Note that audio files have to be smaller than 19.5 MB.

3. Click the green Estimate Price button to see how much it costs to upload the audio file, and then click the green button again to confirm the upload. It may take some time for your audio file to be approved.

4. Once approved, you can access the file from Studio's Toolbox under My Audio (Figure 9.27).

FIGURE 9.27
My Audio.

Summary

In this hour, you've learned about inserting free models and uploading models to Roblox. You save models to Roblox and can share with friends or upload for your own use. You have learned two methods for importing meshes: MeshParts and Asset Manager. If your mesh is ungrouped, you need to use the Asset Manager importing method so that the mesh is ungrouped when it's in Studio. You have learned importing images via Asset Manager and via texture and decal objects. This hour also covered importing audio and sound files. Remember that it takes time for audio files to be uploaded because Roblox checks every file, and they need to follow the Roblox rules. However, sound can add a great value to your game.

Q&A

Q. Does it cost Robux for an audio file to upload on Roblox?

A. Yes, it costs a small amount of Robux to upload the file.

Q. Can I delete models that have been uploaded to Roblox?

A. No, once an object is saved to Roblox, you cannot delete it. However, you can remove public models from your inventory.

Q. Can I edit models once they're saved to Roblox?

A. Yes, but only the information in the Asset Configurations can be changed on an existing model. Only packages can be saved and updated to the newest version in Roblox Studio.

Workshop

Now that you have finished, take a few moments to review to see if you can answer the following questions.

Quiz

1. True or False: You can import 10,000+ triangles mesh via the MeshPart method.

2. True or False: You cannot remove free models from your inventory.

3. To upload audio files to Roblox, you must pay a small amount of _____.

4. It takes audio files some time to get _____.

5. True or False: Meshes cannot be uploaded to Roblox Studio via Asset Manager.

6. True or False: Sounds cannot be uploaded to Roblox via Asset Manager.

Answers

1. False. You cannot import 10,000+ triangles mesh via the MeshPart method.

2. False. You can remove free models from your inventory.

3. You pay a small amount of Robux to upload audio files at Roblox.

4. It takes audio files some time to get approved.

5. False. Meshes can be uploaded to Roblox Studio via Asset Manager.

6. True. Sounds cannot be uploaded to Roblox via Asset Manager.

Exercises

This exercise combines a number of things you've learned in this hour. If you get stuck, don't forget to refer to the previous pages for information! Try to upload meshes from Asset Manager.

1. Open Roblox Studio.

2. Build a forest scene, and rather than creating a lot of different models, bring in a single tree and a single rock.

3. Use one kind of tree and rock. Scale the models so they're different sizes and rotate the models so they're facing different directions.

Your result should look similar to Figure 9.28 when the work is done.

FIGURE 9.28
A finished forest scene.

For the second exercise, team up with a friend and each create one model for the forest scene. Share your models with each other for use in your individual scenes.

HOUR 10
Game Structure and Collaboration

What You'll Learn in This Hour:

▶ How to add places in a game
▶ How to collaborate with others
▶ How to create and access Roblox packages

This hour covers how to structure your Roblox game, including adding places, editing places and scripts, managing collaborators, and creating and using packages. By learning how Roblox games are put together, you can create larger experiences with multiple levels and worlds that players can travel between. Appropriately structuring your game can also improve its functionality. For example, dividing up a large world into multiple places can improve loading time for players.

As your game world becomes more elaborate, you may want to invite a group of people to help you. Using the Team Create and Group functionalities, you can invite collaborators to work with you in real time while sharing assets such as models, scripts, animations, and more.

Adding Places in a Game

Roblox games are made up of individual places; every game has at least one place. One way to conceptualize a place is to think of it as a level in a game. Places contain the environment, models, UI, game logic, and everything else that makes up a level. Figure 10.1 shows a player standing before a selection of places in a game.

FIGURE 10.1
Multiple places of varying themes available for gameplay in "Deathrun Gameshow" by Team Deathrun.

Although a game can be made up of many places, each game has just one starting place that players load into when they start playing.

To add a new place in a game, do the following:

1. In Studio, open the game that you want to add a new place to.

2. Access the View tab and click Asset Manager (Figure 10.2).

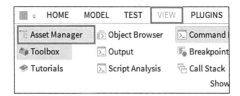

FIGURE 10.2
The Asset Manager.

3. Double-click the Places folder, right-click anywhere in the Asset Manager, and then select Add New Place (Figure 10.3).

FIGURE 10.3
Right-click anywhere in the Asset Manager to add a new place.

You have successfully created a new place. You can double-click it to open it so you can begin editing. Note that the main place of the game always has a Spawn symbol on it in the Asset Manager (Figure 10.4).

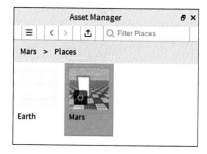

FIGURE 10.4
The Spawn symbol on the main place.

Collaborating in Roblox Studio

Collaborating with other people who bring different experiences, skills, and perspectives can be helpful when creating a Roblox game. A team of people can collaborate on problem-solving, and the ideas and viewpoints from different collaborators can enhance your game's creativity.

While doing so, you should keep in mind that games can belong to an individual or to a group. Most of the games you see on the front page are published to a group because it more easily allows for role management, asset sharing, and managing the Robux earned by a game. However, even if the game is owned by an individual, external collaborators can still be invited to the game to code and create in real time.

Collaborating in Group Games

The Roblox group games feature allows multiple developers to work on the same game, use the same assets, share profits, and give credit to all contributors.

It costs 100 Robux to create a group. Follow these steps to begin:

1. Visit https://www.roblox.com/groups/create and fill in the requested details. (Name and Emblem are required.)

2. Click the green button to create the new group.

Configuring Roles

As a group's owner, you can configure roles for other members in your group as follows:

1. Click the configure button (the ellipsis) in the upper-right corner and select Configure Group (Figure 10.5).

FIGURE 10.5
The Configure menu where Configure Group can be found.

2. In the left column, select the Roles tab (Figure 10.6).

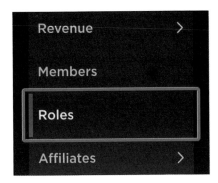

FIGURE 10.6
The Roles option found on the left side of the web page.

3. Once there, note the default roles of Owner, Admin, Member, and Guest. You can change these names/descriptions, and you can create additional roles for 25 Robux. Make sure at least one role other than Owner is allowed to edit.

Assigning Roles

If you're the group owner or have been assigned to a role with the correct permissions by the group owner, you can edit another member's role as follows:

1. In the left column, select the Members tab (Figure 10.7).

Revenue   >

Members

Roles

Affiliates   >

FIGURE 10.7
The Members option found on the left side of the web page.

2. Using the drop-down menu below each group member, select a role.

NOTE

Rules for Changing Roles

You can change the roles of users who are currently in roles ranked lower than yours, and you can promote users to roles ranked lower than your own. To learn more about managing groups, see https://developer.roblox.com/articles/Group-Games.

Enabling Team Create

If the game is owned by an individual, a Team Create session needs to be started. While Team Create is on, invited developers can make changes to the game. To start a Team Create session, do the following:

1. Publish your game by accessing File, Publish to Roblox.

2. Access the View tab and click Team Create (Figure 10.8).

FIGURE 10.8
The Team Create option.

3. Click the Turn On button (Figure 10.9), and Studio starts setting up Team Create.

FIGURE 10.9
Turn on Team Create.

When Team Create is activated, a list of users who are currently active in the Team Create session will appear (Figure 10.10).

FIGURE 10.10
The list of users.

Adding and Managing Users in Team Create

Developers need to be invited to the Team Create session, which you can manage within the Game Settings. Use the following steps:

1. Click Game Settings (Figure 10.11) in the Home tab of Studio.

FIGURE 10.11
Game Settings.

> **2.** Select the Permissions tab (Figure 10.12).

FIGURE 10.12
The Permissions tab.

> **3.** Scroll down until you see Collaborators. Search for an editor by their Roblox username and click their name to add them as a collaborator (Figure 10.13).

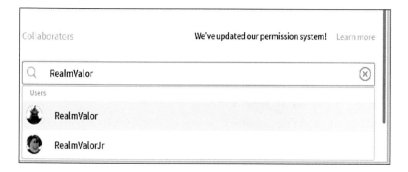

FIGURE 10.13
Searching for Roblox users to add as collaborators.

4. Scroll-down to see the newly added user. On the right, select Edit from the drop-down menu (Figure 10.14) to allow this user to make changes to the game.

Collaborators   We've updated our permission system!   Learn more

RealmValor   **Play**
Can play this game

Users   **Edit**
Can play and edit this game

RealmValor   Play   ×

FIGURE 10.14
Selecting Edit from the Permissions menu.

5. Click Save, and the editor is successfully added to Team Create.

Accessing the Team Create Session

The people who have been invited as a collaborator can join the Team Create session by using the following steps:

1. Open Studio. (If you already have it open, select File, Open from Roblox to reopen it.)

2. Go to My Games (Figure 10.15).

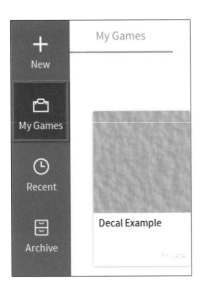

FIGURE 10.15
My Games.

3. Select the Shared With Me (Figure 10.16) tab.

FIGURE 10.16
The Shared With Me tab.

4. Locate the game and start editing.

Using Roblox Studio Chat

Roblox Studio Chat is a tool that allows you to chat as you work with other collaborators. You access it by going to the View tab and clicking Chat (Figure 10.17) to open the Chat window.

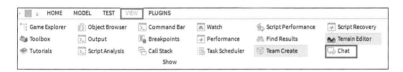

FIGURE 10.17
Opening the Chat window.

Turning Off Team Create

If you're the owner of the Team Create and you have the correct permission, you can turn off Team Create by clicking the ellipsis button (…) on lower-right side of the Team Create window (Figure 10.18). Select Disable Team Create. This action will eject the users in the Team Create Session, and they won't be able to join again until you re-enable Team Create. Previously invited developers will not need to be reinvited.

FIGURE 10.18
Disabling Team Create.

Creating and Accessing Roblox Packages in Roblox Studio

Packages allow you to easily reuse parts of your game, whether it's a model that you want to copy on a level of your game or a collection of scripts that you want to use in multiple games. The Roblox Package allows you to create object hierarchies and reuse them in any game. The advantage to this method rather than doing a copy-paste is that packages are linked to the cloud, so if you make a change to one copy, that change is reflected in all the copies. You can keep packages in sync with the current version by updating any copy within any game. It's like the standard modeling, but in this case, you can update the model any time.

The symbol for a standard model is shown in Figure 10.19. You can easily recognize packages in the Explorer hierarchy by the chain symbol shown in Figure 10.20.

FIGURE 10.19
Standard model.

FIGURE 10.20
Model converted to a package.

Converting Objects to Packages

You can convert reusable objects to a package. You should first group the objects into a model and set the PrimaryPart. If you don't set the PrimaryPart, the model cannot be converted to a package. To convert your asset into packages, do the following:

NOTE

You Can't Delete Packages

You can't delete packages from Roblox because they are assets. However, you can remove them from Studio by accessing Asset Manager, where you click the packages you want to remove in the window that opens. Remove the packages by right-clicking Remove from Game.

1. Select your objects and group them into a model by clicking the Group button (Figure 10.21) in the Home tab and Model tab, or use the Ctrl+G (Cmd+G on Mac) key combination.

FIGURE 10.21
The Group button.

2. Once you have grouped the object, from the Properties window, click PrimaryPart and then select one of the parts in the model (Figure 10.22).

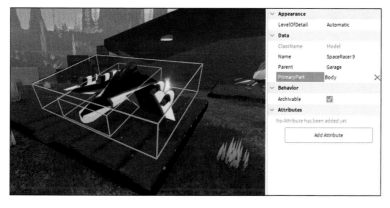

FIGURE 10.22
The PrimaryPart set to the desired part.

3. Right-click the model and select Convert to Package. Fill in the necessary details in the window that opens (Figure 10.23):

  ▶ **About Ownership:** Use Me to retain ownership.

  ▶ **Group:** Save it in a Group by choosing your group name. You see the names of any groups for which you have permission. When you choose a group, the other people who have permission for the group may be able to use the package as well.

4. Click Submit to save the object as a package.

FIGURE 10.23
The Convert to Package window.

Accessing Packages Toolbox

Submitted packages can be found within the Toolbox by clicking the Inventory icon, as shown in Figure 10.24. In the sort menu, select My Packages or Group Packages depending on where you published them in the previous section.

FIGURE 10.24
My Packages in the Toolbox.

Accessing Packages in Asset Manager

Packages used in any games also appear in Asset Manager, which you access from the View tab (Figure 10.25). The Roblox Asset Manager is a useful tab and can be used for many things, such as importing multiple meshes at once, creating new places, and other things.

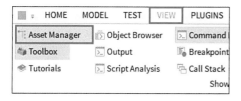

FIGURE 10.25
The Asset Manager.

When the Asset Manager window opens (Figure 10.26), click the Packages folder.

FIGURE 10.26
The Packages folder.

From there, you can insert packages or remove them from the game.

Updating a Package

If you need to change the code included with a package, or update the model, the package can be updated as needed.

1. Once you have made changes to the current package, right-click the model and click Publish Changes to Package (Figure 10.27).

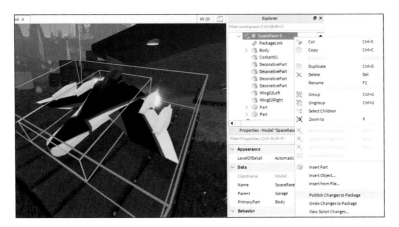

FIGURE 10.27
Publishing changes to a package.

   2. In the prompt dialog box, click Publish (Figure 10.28).

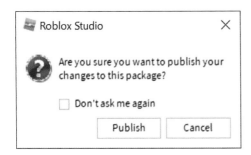

FIGURE 10.28
Confirming that you want to publish.

After pressing Publish, all the changes to the package are successfully saved.

Mass Updating Packages

At some point, you may want to update packages on a game-wide scale. Follow these steps:

   1. Open the Asset Manager window (Figure 10.29). If you have just added the package, close and reopen Asset Manager.

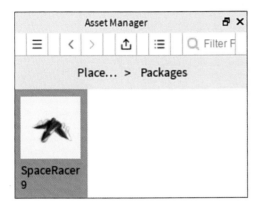

FIGURE 10.29
The Asset Manager window.

2. Right-click the desired package object and select Update All from the context menu (Figure 10.30).

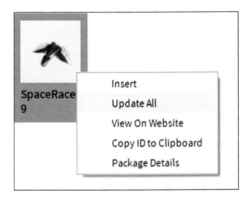

FIGURE 10.30
The Update All option.

3. In the pop-up window (Figure 10.31), select the places within the game to which the mass update should apply and then click the Update button.

NOTE
_____

You Have to Manually Publish Some Changes

This will not autopublish the selected places, so you need to publish them separately if you want the changes to be reflected on the live game servers.
_____

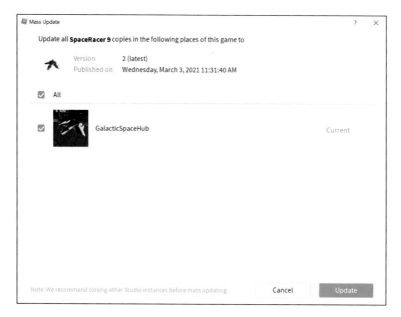

FIGURE 10.31
Selecting the locations for the mass update.

Summary

In this hour, you've learned about creating new places in a game. Each game begins with just one starting place that players load into when they start playing. However, it is possible to spawn in other places in-game by scripting and adding buttons to switch places. You also learned how games can be owned by individuals or by groups. Group members with the correct permissions settings can collaborate in real time. If a game is owned by an individual, collaboration can still occur by creating a session. Finally, you learned about converting objects into packages, accessing them, removing them from the game, and updating the packages. Packages can be used to keep a backup of your game.

Q&A

Q. Can the editors invite more collaborators to Team Create?

A. No, editors do not have permission to invite collaborators to Team Create.

Q. Can I delete my packages?

A. No, but you can remove them from the game.

Workshop

Now that you have finished this hour, it's time to review what you've learned. Take a moment to answer the following questions.

Quiz

1. True or False: Editors can remove the owner of the game from Team Create.

2. True or False: Packages can be deleted from Roblox.

3. True or False: Groups allow for the distribution of Robux among group members.

4. Packages can be easily recognized in the Explorer hierarchy by the _____ _____.

Answers

1. False. Editors can never remove the owner of the game from Team Create.

2. False. Packages cannot be deleted from Roblox; however, they can be updated.

3. True. This is one major advantage they have over individually owned games.

4. Packages can be easily recognized in the Explorer hierarchy by the chain symbol.

Exercise

To this point, your game has consisted of a single place. Take time and plan out how your game can be expanded to include multiple places. Here are a few examples:

▶ A massive multiplayer fantasy world where home areas for each race are created in a different place file.

▶ A competitive shooter where players can select a map for each round.

▶ A game set in outer space where players unlock new worlds as they level up.

Hour 17, "Combat, Teleporting, Data Stores," explains how to teleport players between places. In the meantime, write down your plans and begin thinking about how much time you want to spend creating each place and if you'll want to collaborate with friends using a group.

HOUR 11
Lua Overview

What You'll Learn in This Hour:

▶ How to use the coding workspace
▶ How to use variables to modify properties
▶ How to work with functions and events
▶ How to set up conditional statements
▶ What arrays and dictionaries are
▶ How to use loops to execute code multiple times
▶ How to access variables and functions within scope
▶ How to create custom events
▶ How to find mistakes through debugging

Roblox could choose from many coding languages when deciding what would work best for its environment, and Lua is the language of choice. Lua requires fewer words than most programming languages (such as Java and C++), which makes it easier to read and faster to type. Roblox Lua is a modified version of Luae.

This hour give you a brief overview of coding terminology and Lua scripting before you delve into more advanced game development topics—such as GUIs, animation, and camera movements—that the remainder of the book covers. You'll create and modify a part via different types of scripts and receive an introduction to object-oriented programming. Finally, you'll run your scripts to see how to read the output and debug errors.

Using the Coding Workspace

Programming languages are made of lines of code. Sets of instructions that have been put together are called a script, which is just a way of telling the computer what needs to be done. By using Lua scripts in Roblox Studio, you give it instructions to create 3D elements in your game and make them interactive, like responding to an input from the player or a series of events that need to happen in a gameplay.

Before you start coding, make sure your workspace is arranged in an optimal way to test your script and see its output. You need the output window for the tasks in this hour, so go ahead and enable Output under the View tab. Figure 11.1 is what your workspace arrangement should look like.

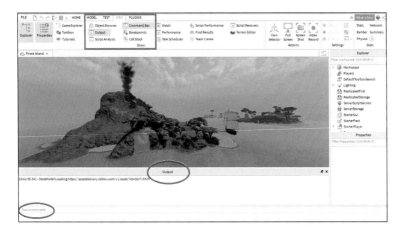

FIGURE 11.1
The workspace arrangement with the output window beneath the 3D game space editor.

Creating Your First Script

Add a new Script object to ServerScriptService that will be used to hold your code. Locate Server-ScriptService and use the plus sign to insert a script.

The script editor automatically opens on a separate tab next to your game world (Figure 11.2).

FIGURE 11.2
The open editor on the tab next to your game world.

Whenever you insert a new script, a `print("Hello World!")` statement is always present by default (refer to Figure 11.2). The `print` function displays whatever message is inside its parentheses—in this case, "`Hello World!`"—in the output window when you run the program.

Rename the script **HelloWorld** so that you can identify it later. You can rename a script the same way you rename a part: double-click or right-click it in Explorer and enter a new name in the Properties name field.

Now run your `HelloWorld` script by clicking on the Play button on the top ribbon. You see `Hello World!` printed in the output window (Figure 11.3).

FIGURE 11.3
Playtesting the script.

Using Variables to Modify Properties

You can use code to print messages into the output window, but it also has many other uses as well. For example, you can use scripting to modify an object's property. Normally, you would modify properties simply by updating the relevant field in the Property window. However, the advantage to modifying a property via scripting is that the change will occur *when the game runs.* This allows you to have dynamic changes throughout your game, making it a much more exciting player experience. Let's look at what variables are and how they work to help modify properties.

Overview of Variables

Variables are like containers where information—values such as numbers, strings, booleans, data types, and more—can be maintained and referenced. (Data types are the different types of data a variable can store. For reference on the different types of data, see Appendix A, "Lua

Scripting References.") The values can change, but the program remains the same, which is why it's possible for the same program to process different sets of data.

When you're creating and naming variables, make sure to choose names that represent the information the variables are going to contain. Careful naming increases your code's readability. Here are some guidelines for creating variable names:

▶ Variable names are case-sensitive, which means that *RedBrick* and *REDBRICK* are two unique names.

▶ Variables cannot be reserved keywords—for example, `if`, `else`, `and`, `or`, and so on.

▶ Variable names can be of any length and consist of letters, digits, and underscores, but they *cannot start with a digit*.

▶ You cannot use spaces in a variable name.

▶ You cannot use special characters except the underscore.

▶ Don't give a variable the same name as a function because the variable will override the function.

▶ Avoid names that begin with an underscore followed by uppercase letters (such as `_VERSION`) because they may be reserved for internal global Lua variables.

▶ When trying to reference parts in a script, make sure you have distinct names for them so as to not confuse the program.

Creating Variables

You assign a value to a variable with the equal sign (=) operator. The variable is always on the left of the equal sign, and the value is on the right. The variable can be preceded by its scope, which is the location where a variable can be "seen" and accessed. (Read more on scope later in this hour.) A variable can be diagrammed as shown in Figure 11.4.

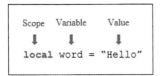

FIGURE 11.4
Diagram of a variable.

With the script `print (word)` shown in the figure, your output window would display `Hello`. Once declared, a variable's value can be changed by simply assigning another value to it.

Create a Translucent Bomb

Let's say you have a bomb that you want to be translucent when the game starts. You need to change the transparency property of the bomb to 0.5 (1 being completely transparent and 0 being completely visible):

1. Insert a part in your workspace. Rename it **Bomb**.

2. Access the Bomb via code so we can change its property. The objects are nested in a folder structure (Figure 11.5). Game is the top object (not shown in the Explorer window), then Workspace, and the part to modify underneath.

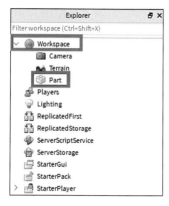

FIGURE 11.5
Hierarchical representation of the elements in the game world.

3. To tell the computer exactly where to find the part to modify, provide its location separated by periods:

```
game.Workspace.Bomb
```

4. To change the part's transparency and make it translucent (0.5) when the game starts, you can use a variable as follows:

```
local translucentBomb = game.Workspace.Bomb
```

5. Add the property you want to change—once again, separated by a period—and assign the value you want:

```
translucentBomb.Transparency = 0.5
```

Adding Comments to Your Code

Comments are text added by the programmer to explain the code's intent; they're bits of explanation that aren't executed when you run the program. It's like leaving a note for yourself and anyone else who reads your code.

There are two types of comments:

▶ **Single line or short comment:** Starts with a double hyphen (--) anywhere on the line and extends to the end of the line, like this:

```
-- This is a comment
local var = 32 -- This is a comment following code on the same line
```

▶ **Multiline or block comments:** Starts with --[[and ends with]]--, like this:

```
--[[
This is a long comment.
It can contain short comment hyphens, like this:
--
--
--]]
```

▼ TRY IT YOURSELF

Writing Comments

Earlier, you wrote code to modify a part's transparency. Now modify the part's reflectance. Write a comment for your code describing why you needed it to be reflective.

Create another bomb under the workspace. You can copy/paste the existing one and move it around a bit because it will be created over your first bomb. Now there are two bombs under the same hierarchy (game.Workspace.Bomb), so how will the ServerScriptService script be able to identify which part should be translucent?

For the script to discriminate between bombs, you can give each one a unique name. Rename the first part TranslucentBomb and modify the script as follows:

```
-- This script changes the transparency of TranslucentBomb

-- Create a variable to store the bomb
local translucentBomb = game.Workspace.TranslucentBomb
translucentBomb.Transparency = 0.5
```

What if you have multiple bombs that need to be translucent? Instead of following these same steps each time—making new bombs and modifying them in ServerScriptService script—you can attach each bomb with its own script instead of calling the number of bombs out from the script in ServerScriptService.

You can make the script you just wrote reusable. Within the workspace, you can make use of the parent-child relationship, where the part is the parent and the child is the script. In Figure 11.6, the parent is `TranslucentBomb` and the child is the script.

FIGURE 11.6
Hierarchical representation of the elements of your game world. Here the Script objects are each parented to a bomb part.

To make the script reusable, instead of using `game.Workspace.TranslucentBomb`, you can use `script.Parent` to tell the script to look for the parent (in this case, `TranslucentBomb`).

```
-- Create a variable to store the bomb
local translucentBomb = script.Parent
translucentBomb.Transparency = 0.5
```

This way, the code affects the script's parent no matter what it's named.

Using Functions and Events

A function is a sequence of instructions packed as a unit to perform a certain task. In the original script example, `print("Hello World!")`, `print()` is a function that allows messages to be displayed in the output window. The words inside the parentheses are called the *argument*. Arguments are the information being passed along for the function to use—in this case, "Hello World!"

Once defined, you can use (or call) a function multiple times as a command or trigger it through an event. Functions are usually named in camelCase, where the first letter of the first word is lowercase and the first letter of the second word is capitalized. There are no intervening spaces or punctuations.

Creating a Function

A function is defined with the keyword `local`, then the keyword `function`, followed by the name of the function in camelCase, and then the parentheses with no space between them, like so:

```
local function nameOfTheFunction()
-- Your code indented here
end
```

The function body is where your logic or code exists. This needs to be indented inside, after which it must use the keyword `end` to end the function definition. Now that the function is defined, you can use or call it multiple times as a command or triggered through an event.

To call a function, type its name followed by parentheses, as shown here:

```
nameOfTheFunction()
```

Using a Function to Explode a Bomb

Write a function in ServerScriptService to explode the part bomb that was previously created. Explosions exist as objects in Roblox already, so you create an instance, or a copy, of the Explosion object using code. Copy the following code into the script editor:

```
local explodingPart = workspace.Bomb

-- This is a function to explode a bomb
local function explodeBomb(part)
        -- Code needs to be indented
        local explosion = Instance.new("Explosion") -- Creates an explosion
        explosion.Position = part.Position print("Exploding") -- Helps check if the
                function ran.
        explosion.Parent = explodingPart -- adds it to the workspace.
end
wait(7) -- Function for making the code wait
-- Function call
explodeBomb(explodingPart) -- Sends the part as an argument
```

To see properties of Explosion that you can modify, go to https://developer.roblox.com/en-us/api-reference/class/Explosion.

Destroying the Bomb

Now you can playtest your code to see the bomb exploding. You can see the explosion, but the part remains intact and does not get destroyed. Write a function to destroy the bomb after it explodes; you can write the function in the same script. Read more about it at https://developer. roblox.com/en-us/api-reference/function/Instance/destroy.

Solution:

```
local function destroyBomb(part)
     print("This part is Destroyed")
     part:Destroy()
     wait(1)
     part.Parent = game.workspace
end
destroyBomb(explodingPart)
```

Using Events

In addition to properties and functions, objects also have events. An event is a signal that is fired when something important happens. For example, when a player touches a part, a Touched event is fired. Other functions listen for the event to fire before running their code.

This allows for cause-and-effect systems to be set up. For example, when a player scores a goal, an event named Score might be triggered. Functions listening for the Score event would then know to run the appropriate code to update the scoreboard.

Using an Event to Explode a Part When Touched

Previously, we had the functions to explode and destroy the bomb. Now you add the following new function that is called whenever the Touched event is fired. Within it, you call the two previously coded functions:

```
local function onTouch(obj) -- The obj that is passed to the function here is the
     player.
     if obj.Parent and game.Players:GetPlayerFromCharacter(obj.Parent) then
          explodeBomb(explodingPart) -- we are sending the part as a parameter
          destroyBomb(explodingPart)
     end
end
--Touched event connects to the onTouch function
explodingPart.Touched:connect(onTouch)
```

The line of code at the bottom connects the function named onTouch to the bomb's Touched event. This way, onTouch runs whenever Touched is fired. When Touched is fired, it includes the name of whatever touched the part in the signal.

Next, you check if what touched it is a player using the Players:GetPlayerFromCharacter method of the players. (See the next section for more about conditional if structures.) After that, you call the explodeBomb and destroyBomb functions.

NOTE

Roblox Class API

To know which functions and events to use, refer to the Roblox Class API at https://developer.roblox.com/en-us/api-reference/index.

Working with Conditional Statements

Conditional statements allow scripts to perform actions when specific conditions are met. For example, *if* the player is out of health, *then* end the game. If a condition is met, Lua treats it as true. If the condition is not met, the value is either false or nil. These conditionals can be checked by using the relational operators, which you can review in Appendix A.

The if block is used to specify a block of code to be executed *only if* the specified condition is true.

For example, you can create another bomb and name it ColorBomb, set its color to blue, and then change the color from the property window to gray, just to see how the else block works:

```
local colorBomb = script.Parent
if (colorBomb.BrickColor.Name == "Really blue") then
      print(" ColorBomb is blue")
end
```

The else block is used to specify a block of code to be executed when the if condition is false:

```
if (colorBomb.BrickColor.Name == "Really blue") then
      print("ColorBomb is blue")
else
      print("ColorBomb is " .. colorbomb.BrickColor.Name)
end
```

elseif is used to specify a new condition to test, if the first condition is false. There can be as many elseif tests as possible, and it is executed in order of occurrence:

```
if (colorBomb.BrickColor.Name == "Really blue") then
    print("ColorBomb is blue")
elseif (colorBomb.BrickColor.Name == "Dark stone grey") then
    print("ColorBomb is grey ")
else
    print("ColorBomb is " .. colorBomb.BrickColor.Name)
end
```

Understanding Arrays and Dictionaries

There's a lot of information within games that needs to be tracked—things like scores, inventory items, and who is on what team. This information is typically tracked within data structures called tables. A *table* is a data type that stores different values like numbers, booleans, strings, and even functions. Tables can behave as arrays or dictionaries, and empty tables can be created by using curly braces, as shown here:

```
local playersBeingWatched = {}
```

An *array* is a table with a list of ordered values. You can access the value in the order in which it occurs, starting from 1. In other words, arrays have numbered indexes. Say we have an array of players:

```
local playersBeingWatched = {'Player1', 'Player2', 'Player3'}
print(playersBeingWatched[1]) -- this will print out Player1
```

A *dictionary* is a table of key-value pairs, where keys are used to identify the values instead of numbered indexes. You use dictionaries when you want to label your values. Say we want to store a player's age:

```
local playersAge = {
    Player1 = 16,
    Player2 = 15,
    Player3 = 10
}
print(playersAge[Player2]) --this will print out 15
playersAge[Player5] = 20 --this will add to the existing dictionary
```

Using Loops

A loop lets you execute the same, or similar, code multiple times, which is useful because it allows you to write one set of instructions that operates on multiple separate sets of data. You might need every player within an array to be placed onto a team or an animation to play over and over. Lua includes a few types of loops, and they each repeat blocks of code in different ways.

while **loop**

With the `while` loop, you can execute a single statement or a block of statements as long as a condition is true. The `while` loop might not be executed at all if the condition stays false. When the condition is tested and the result is false, the loop body is skipped, and the first statement after the `while` loop will be executed. Here's the syntax for the loop:

```
while(condition)
do
    statement(s)
end
```

Create another bomb and name it **colorSwitchingBomb**. You want this bomb to change between two different colors and materials, and we want this to happen throughout the game, so the condition in this case is going to be `true` always:

```
local colorSwitchingBomb = script.Parent
while true do
    colorSwitchingBomb.BrickColor = BrickColor.new("Bright blue")
    colorSwitchingBomb.Material = ("Neon")
    wait(1)
    colorSwitchingBomb.BrickColor = BrickColor.new("Bright red")
    colorSwitchingBomb.Material = ("SmoothPlastic")
    wait(1)
end
```

wait()

A commonly used function is `wait()`, which you use to make the script pause for a certain amount of time. Here are some reasons you might need to pause the current process:

▶ Sometimes the changes you expect to see happen within a fraction of a second, which is not visible to your eyes. In such cases, you need to add the `wait()` function with the appropriate wait time.

▶ When you experience a lag, there are objects without the `wait()` function that use up compute time. The solution for this is to include a wait for lower-priority events. (See the next section for more about events.) In other words, you add pauses until the event occurs.

It is recommended that you always include a time value argument when using `wait()`. If no time is supplied, then it typically returns in 0.03 seconds. We talk about `wait()` more as we proceed through the hours.

repeat-until **loop**

A repeat-until statement repeats its body until its condition is *true*. The conditional expression appears at the end of the loop, so the statements in the loop execute once before the condition is tested. If the condition is false, the flow of control jumps back up to the top of the instructions, and the loop executes again and again until the given condition becomes true. Here's the syntax for a repeat-until loop:

```
repeat
      --code
until( condition )
```

Say you want your part to change color and material only if the BrickColor is not already Bright blue. The code would look like this:

```
local colorSwitchingBomb = script.Parent
count=0
repeat
      colorSwitchingBomb.BrickColor = BrickColor.new("Bright blue")
      colorSwitchingBomb.Material = ("Neon")
      wait(1)
      colorSwitchingBomb.BrickColor = BrickColor.new("Bright red")
      colorSwitchingBomb.Material = ("SmoothPlastic")
      wait(1)
count = count +1
until(count=6)
```

for **loop**

The for statement has two variants: the numeric for and the generic for. A numeric for uses three values to control how many times they run: a control variable, an end value, and an increment value (Figure 11.7). Starting from the value of the control variable, the for loop either counts up or down each time it runs code inside the loop until it passes the end value. Positive increment values count up; negative increment values count down.

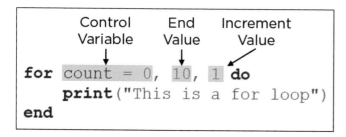

FIGURE 11.7
The three values in a numeric for.

The syntax looks like this:

```
for count = 10, 0, -1 do
    -- Prints the current number the for loop is on
    print(count)
    -- Wait 1 second
    wait(1)
end
```

ipairs() **and** pairs()

Commonly, you'll want the same thing to happen to every object in an array or table. You use ipairs() to repeat code for every object in an array and pairs() to repeat dictionaries.

▶ ipairs(): For each step in the code, this works with the index and value:

```
-- print all values of array 'a'
local playersBeingWatched = {'Player1', 'Player2', 'Player3'}
for index, value in ipairs(playersBeingWatched) do
    print(index,value)
end
```

▶ pairs() can be used to work with a dictionary element's key, value, or both:

```
-- print all values of dictionary 'playersAge'
local playersAge = {
    Player1 = 16,
    Player2 = 15,
    Player3 = 10
        }

for key, value in pairs(playersAge) do
    print(key,value)
end
```

Working with Scope

Not all variables or functions can be accessed from anywhere in the program. The part of the program where a variable or a function is accessible is called its *scope*.

▶ **Local scope:** When a variable or function is prefixed by the local keyword, they become locally scoped, which means it is only available within that function. When you define a variable within a function, it is accessible from the point at which it is defined until the end of the function, and it exists as long as the function is running. Local scoped values cannot be accessed or changed from outside the function. In the local function explodeBomb defined earlier, variable local explosion is accessible only inside the function explodeBomb and not outside it. And the local function explodeBomb is accessible only inside the script and not outside that script.

▶ **Global scope:** Once declared, global variables and functions can be used by any following chunk of code in the same script. Variables and functions default to global scope unless they are marked with the local keyword.

▶ **Enclosing scope:** Loops, functions, and conditional statements all create a new scope block. Each block can access local variables/functions in its parent block, but not those in any child blocks.

TRY IT YOURSELF ▼

Creating Local and Global Variables

You've already seen so many examples using local variables. Create local and global variables and assign them to a string. Try accessing them outside their scope and see if they can be accessed.

Creating Custom Events

Roblox provides a number of already created events, such as Touched, which was demonstrated earlier in the hour. To create your own, use BindableEvent.

There are times when you need to customize the event for your game, like signaling the start of a match, beginning the timer, signaling the end, and stopping the timer.

BindableEvent allows events defined in a script to be subscribed (or connected) by another script of the same scope. The event BindableEvent.Event:Connect() defined in the Event-Subscriber is fired using BindableEvent:Fire() in the EventPublisher. Multiple scripts can listen for the same bindable event, which helps keep code organized and easier to modify.

In the previous event example, when the player touched the bomb, it exploded and was destroyed. After it exploded, we wanted to destroy the bomb; these are two series events (which is custom in our case) that we want to make happen every time a bomb is touched. (The game world wouldn't just have one bomb, right?) You can use BindableEvent in this case:

1. Create a bindable event in the workspace.

2. Create a script EventSubscriber under ServerScriptStorage. This script connects to your custom event on fire.

3. Create a script EventPublisher under the bomb. This script fires the bindable event onTouch and sends the bomb as a parameter:

▶ EventSubscriber

```
local BindableEvent = game.workspace.BindableEvent
local function explodeBomb(part)
local explosion = Instance.new("Explosion") -- to create an explosion,
create a new instance
        explosion.BlastRadius = 15 -- damage area distance
        explosion.Position = part.Position -- explosion happens at part
            location
        print("Exploding") -- Print for debugging purposes, in case the
            blast happened before we could see it.
        explosion.Parent = game.Workspace -- the parent property of the
            part is locked,
end
local function destroyBomb(part)
        print("This part is Destroyed")
        part:Destroy()
        wait(1)
        part = nil -- After destroying an object, set its descendants to
            nil
        part.Parent = game.Workspace
end
function customevent(child)
        --Any code in here will run when the Bindable event is fired
        explodeBomb(child)
        destroyBomb(child)
        print("inside our custom event")
end
--listening event which is always on the lookout
--This event triggers when the Fire method is used
be.Event:Connect(customevent) -- the above function will now fire when
        the event fires.
```

▶ EventPublisher

```
local explodingPart = script.parent -- Variable declared before the
        function
local be = game.Workspace.BindableEvent
local function onTouch(obj) -- The obj that is passed to the functions is
        the player.
if obj.Parent and game.Players:GetPlayerFromCharacter(obj.Parent) then
        --Fire is used to trigger the event
        be:Fire(explodingPart) -- we can pass in an argument here
        print("Event firing")
        end
end
explodingPart.Touched:connect(onTouch)
```

Debugging Code

When you're developing a game, you inevitably will need to debug it for errors because it's not always possible to write perfect scripts or figure out what's wrong in the program. For instance, the code may be syntactically correct, but it's not functioning as required. Being able to debug is an important skill of a good developer. Debugging and testing are complementary processes. When you test (also called playtest), you find out any errors; the point of debugging (traceback) is to locate and fix the mistake. Roblox has provided some helpful debugging tools to catch bugs easily.

String Debugging

When you playtest the game, the output window displays user-defined messages (print) and errors from running scripts. Using print at key places of your script helps you debug code when accompanied by defined messages. For example, look at the "Using Functions and Events" section earlier in the hour where we used print statements to make sure a function is being called.

Lua Debugger

The Lua debugger enables you to debug code using breakpoints. Breakpoints are nothing but places where the game pauses and runs step by step. The Lua debugger is enabled by default in the Settings (Figure 11.8). You can toggle it on or off.

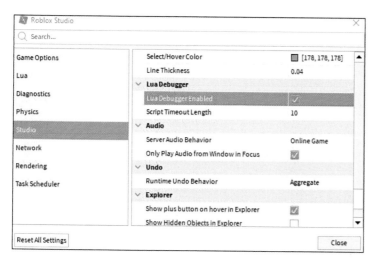

FIGURE 11.8
The Lua debugger.

Say your game world has numerous scripts, and the game breaks during playtesting. When you're not sure what's breaking your game, you can use the Lua debugger:

1. Create a breakpoint in your script.

2. Left-click the line number of the code, and select the Breakpoint option. You're given an option to Insert Breakpoint.

3. When you click the option, a red dot follows the line number, showing the breakpoint.

Now, when you test the game, it pauses at your breakpoints one by one. You can control and inspect the breakpoints in the breakpoint window, which you can enable from the View tab. Once the error is debugged, you can delete the breakpoint by clicking the red dot or by right-clicking the breakpoint. You can also disable it the same way, in case you want to repeat the breakpoint.

Log Files

A log file keeps track of everything that happened from the minute you start running until you stop. The log files are automatically created to store error or warning messages. They are located in a folder created locally on your desktop, so you don't need your application to run to view logs. On Windows, the log file is under %LOCALAPPDATA%\Roblox\logs; on a Mac, it's under ~/Library/Logs/Roblox.

You look at the log files when you want to see an older session's errors and probably want to check if the same errors have been repeating. (Sometimes you might have performance issues based on one error that's been repeating.) All log files are stored in the format log_XXXXX, followed by additional naming. All logs with the same XXXXX value are from the same Studio session.

There also may be situations when a Roblox Customer Care employee requests these log files to investigate issues, or you might want to post it on the Dev Forum too.

▼ TRY IT YOURSELF

Practicing Debugging Skills

Try to include errors on purpose to see what kind of errors or warnings are created by Lua. Use the print statements and the Lua debugger for this. By doing this, you are polishing your debugging skills to help you more easily identify bugs in the future, thus making you a better developer.

TIP
_____

Becoming a Better Game Developer

Scripting in Lua can make your game world interactive, and you'll be impressed by its impact. There are a few things to consider:

▶ **Plan:** Always remember to plan your game strategy and its functions. Create flowcharts to understand the flow. You might then have questions such as, "Does this piece of code need to be a script at the server or a LocalScript at the client, or maybe one must create a ModuleScript for this?" You will be able to answer all these questions and decide which is best for you.

▶ **Try:** Always put action to your ideas. When you start implementing your ideas, you might have setbacks or better ideas, but keep going.

▶ **Code can always be better:** There are different ways to code the same functionality. Starting with the basics will help you understand the vast classes in Roblox.

▶ **Divide and conquer:** Divide your ideas into different templates. Keeping all the parts and scripts in one template might get confusing when the game isn't working as you anticipated. You can test combinations of these parts by creating models.

_____

Summary

In this hour, you learned how Lua is used in Roblox and how to organize your coding workspace. You learned how to code scripts and make them reusable by using the parent-child relationship. You also learned how to modify simple properties of objects. You looked at the different kinds of loops and conditional structures and how code is scoped locally and globally. You learned to make custom events and bindable events and how to debug code.

Q&A

Q. What is scripting, why is it important, and what language does Roblox use for scripting?

A. Scripts contain a set of instructions that tell the computer what needs to be done. Scripting is necessary for game developers; to make your game completely interactive, you need to know scripting. Roblox uses a language called Roblox Lua.

Q. What is the use of a function and an event?

A. Functions are sets of instructions that can be used multiple times in a script. Once defined, a function can be executed through a command or triggered through an event.

Q. What is the use of a `pairs()` and `ipairs()`?

A. `ipairs()` is used to iterate or repeat through an array and `pairs()` is used to iterate or repeat through a dictionary.

Q. When do we use log files versus the Lua debugger?

A. You can use the Lua debugger when you are not able to locate the error caused by a piece of code, and the log files when your studio isn't starting up or you want to look at all the older log files.

Workshop

Now that you have finished, let's review what you've learned. Take a moment to answer the following questions.

Quiz

1. What is the optimal arrangement for your scripting workspace?

2. True or False: Comments are required in every line of code.

3. True or False: Bindable events are used to connect the server and the client.

Answers

1. Closing the extra windows gives you more space to see what you're doing. It keeps the Explorer and properties windows aligned below each other (look at Hour 2 if you're not able to align them one below the other) with the output window open at the footer of the editor.

2. False. Comments are like little messages to yourself or the code viewer to understand the code or logic better.

3. False. Bindable events are used to connect a client to another client or a server to another server.

Exercise

This exercise combines a number of different things you've learned in this hour. If you get stuck, don't forget to refer back to the previous pages in this hour!

Create a bomb with three explosions. You want to be able to copy/paste these triple blast bombs in multiple places in your game world. To do this, you need to create three bombs—one on top of the other—and make the top two invisible. This way, the explosions happen three times but vertically. You can use the same EventSubscriber, but you need to modify the EventPublisher to make a note of any child bombs and explode them first.

Solution:

```
local explodingPart = script.parent -- Variable declared before the function
local be = game.workspace.BindableEvent
local function onTouch(obj) -- The obj that is passed to the functions is the
    player.
```

```
    if obj.Parent and game.Players:GetPlayerFromCharacter(obj.Parent) then
        --Fire is used to trigger the event
        local children = workspace.Bomb:GetChildren()
        for i, child in ipairs(children) do
            local child = children[i]
            if(child.Name == 'Bomb') then
                print(child.Name .. " is child number " .. i)
                be:Fire(child)
                print("Event Firing")
            end
    end
    be:Fire(explodingPart) -- we can pass in an argument here
    print("Event firing")
    end
end
explodingPart.Touched:connect(onTouch)
```

HOUR 12
Collisions, Humanoids, Score

What You'll Learn in This Hour:

▶ What collisions are

▶ How to detect collisions

▶ How to work with Humanoids

In Hour 4, we introduced you to Roblox's Physics Engine, which handles how physical objects move and react. In that hour, we also briefly discussed collisions, but here we go into more depth on how collisions are handled with more complex objects, such as meshes, unions, and groups. We also introduce Humanoids, the special objects that give models the functionality of a character, and we show you how to create a realistic walking Humanoid.

Introduction to Collisions

You may recall from Hour 2, "Using Studio," that collisions happen when two objects (or rigid bodies) intersect or get within a certain range of each other. In Roblox Studio, the Collisions toggle enables and disables collisions for the purpose of editing. When collisions are on, you can't move a part into any place where it overlaps another part. However, it does not affect whether items in game are collidable. As you move parts and basic objects, you may notice a white outline whenever a part touches another part. This indicates that a collision is happening. This collision box is the indicator for these simple objects, but for more complex objects, such as imported meshes and unions, the CollisionFidelity property exists.

CollisionFidelity

The CollisionFidelity property is used for finding the sweet spot between performance and accuracy both while editing and in game. The more detailed the collision box, the more costly it is in terms of performance. As such, developers often disable a parts collision and use invisible parts for collision and collision detection.

A good example of developers removing complex mesh collisions is in fast-paced action games where they may place an invisible object over different areas of the player's character to detect bullets and so on.

Figure 12.1 shows CollisionFidelity's properties.

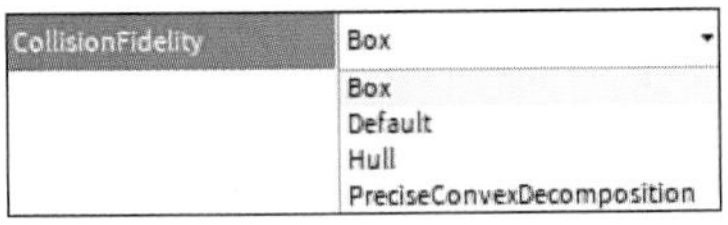

FIGURE 12.1
CollisonFidelity options.

Showing and Improving Collision Geometry

If you try changing the CollisionFidelity property on a mesh or a union, you may notice it's difficult to tell what impact each CollisionFidelity option makes on the object. To give the option a more visible difference, use the following steps:

1. From the File menu, select Settings and then Physics.

2. Turn on Show Decomposition Geometry.

3. Restart Studio.

When Show Decomposition Geometry is turned on, your object should change color as long as your object is a mesh or union (Figure 12.2).

In Figure 12.2, the default CollisionFidelity is inaccurate, causing the player or objects to walk seemingly in the air. This is because Roblox is attempting to calculate collisions while remaining as performant as possible. The more complex the collision is, the more computation it requires.

To correct this, you should use CollisionFidelity to see which setting offers the most accurate geometry without being too costly to your performance. In this particular case, it required the PreciseConvexDecomposition setting, which is quite costly to performance.

TIP

Maximizing Performance

Try to keep the number of vertices in your meshes low—for collidable objects, specifically, but also in general—to be performant. Higher poly objects require more computationally intensive calculations.

PreciseConvexDecomposition is a performance-taxing algorithm that's also unoptimized like many unions. It may be much more complex than necessary.

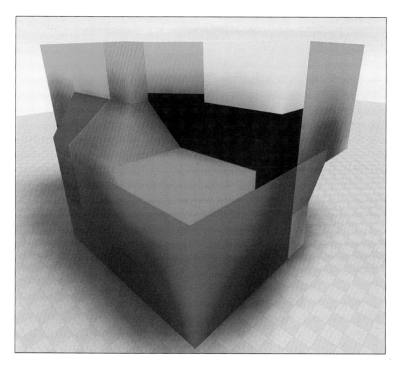

FIGURE 12.2
Union with Decomposition Geometry visible.

Working in the Collision Groups Editor

There are two properties tied to collisions on a regular part: CanCollide, which we discuss in Hour 4, "Building with Physics," and Collision Groups. Collision groups are a way to group objects and control whether they can collide with objects in other groups. The Collision Groups Editor is a method of directly modifying collision groups. It's perfect for adding or removing collision groups and modifying how they interact.

You use a table system (Figure 12.3) in the Collision Groups Editor to manipulate how different collision groups interact with each other. Where the row and column meet, you can pick whether the groups collide. With this system, you have a simple method to build complicated collision behavior and layers. Collision groups can be created and edited either directly in the editor or by code using related APIs.

Open the Collision Groups Editor by selecting the Model tab and navigating to Collision Groups within the Advanced section (Figure 12.4).

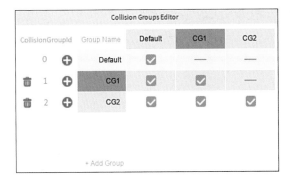

FIGURE 12.3
Collision Groups Editor with CollisionGroupId visualization.

FIGURE 12.4
Collision Groups under the Advanced section.

Using the Collision Groups Editor Manually

The Collision Groups Editor supports four basic but powerful functions:

▶ Editing how two groups interact on all but the default layer

▶ Adding layers

▶ Renaming layers

▶ Deleting layers

Collision groups can be assigned to different parts by simply selecting the part and clicking the preferred collision group:

▶ **To add a new collision group:** Enter the name into the Add Group field at the bottom of the editor and press Enter.

▶ **To remove a collision group:** Click the bin symbol next to the collision group.

▶ **To add an object to a collision group:** Select the object and click the plus button.

▶ **To edit a collision group name:** Click the notepad button next to the collision group.

▶ **To edit how collision groups interact:** Select the two groups you want to edit. Find where the column and row intersect. Check or uncheck the field.

Using the Collision Groups Editor via Script

You also can modify collision groups via script through the Physics Service using the following code:

```
PhysicsService:GetCollisionGroupId("CollisionGroupName")
PhysicsService:GetCollisionGroupName(CollisionGroupId)
PhysicsService:CreateCollisionGroup("string")
PhysicsService:SetPartCollisionGroup(workspace.Part, "CollisionGroupName")
```

This enables you to modify collisions in game—that is, in real time via code—which opens many possibilities and uses, such as trap doors, VIP entrances, and non-collidable players between team members. In the following exercises, you work through a couple of examples so you better understand how they work.

TRY IT YOURSELF ▼

Create a Collision Group

Let's test out the Collision Groups Editor quickly before testing the API:

1. Place down a part.

2. Make sure the object is anchored.

3. Make sure CanCollide is set to true.

4. Make sure your Collision Groups Editor is open.

5. Create a new collision group and set the block to that collision group.

6. Find where the default collision group and your new layer overlap and deselect it, as shown in Figure 12.5.

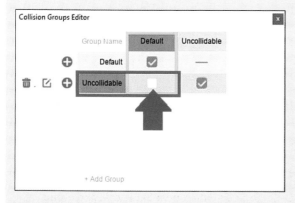

FIGURE 12.5
Test 1 visualization.

▼ TRY IT YOURSELF

Use the Physics Service to Switch Collision Groups

Test out the Physics Service API quickly before you move on to detecting collisions. You're going to use code with the API to revert this block to the default collision group. Use the following steps:

1. Create a new script within the part.

2. Call the Physics Service, setting it to a variable as you do.

```
local PhysicsService = game:GetService("PhysicsService")
```

3. This gives you access to the relevant functions so you can reference :SetPartCollisionGroup(), with the part instance and the name of the requested collision group as parameters.

```
local PhysicsService = game:GetService("PhysicsService")
PhysicsService:SetPartCollisionGroup(script.Parent,"Default")
```

NOTE

Setting CollisionGroupID to 0

Alternatively, you could set the parts CollisionGroupId property back to 0 (the default) without even referencing the Physics Service, but this is not recommended.

Detecting Collisions

.Touched is the native collision detector, and it fires every time two objects collide. This is called a .Touched event. It's used all the time, especially when creating traps, collecting coins, and creating buttons. This allows developers to easily detect when, how, and where the player is in the world.

Using .Touched

.Touched is natively built-in, which means it's really easy to use because there's very little extra work we need to do in terms of detection. Simply reference .Touched, hook the event to a function using :Connect, and voilà. Here's an example:

```
local detector = script.Parent
local function partTouched(part)
      print("Touched: "..part.Name)
      wait(1)
      print("Complete")
end
Detector.Touched:Connect(PartTouched)
```

.Touched is perfect in some cases, but you should keep in mind some things mentioned in Hour 11. The biggest issue with .Touched is its sensitivity; it often fires multiple times if you move while touching an object. We often use other methods to work around this, including the following:

- ▶ Debounce

- ▶ Magnitude, Distance Checking

- ▶ Region3

- ▶ Raycasting

- ▶ GetTouchingPart()

The following sections cover .Touched and the common practices used to make it function in a reliable fashion, as well as some use case examples.

Debounce

Debounce is a tool to limit the number of times a function will run. It is mainly used in conjunction with .Touched because .Touched often fires multiple times in short succession for the same object. You can use a debounce to keep .Touched in check and to tell if the function's already running. Using a script, we can see the output both with and without debounce:

Output without debounce:

```
Touched: Part
Touched: Part
Touched: Part
(wait)
Complete
Complete
Complete
```

To set up a debounce, you can do the following:

1. Set up a local variable (a boolean) outside of the function so it's not accidentally redeclared.

2. Check whether the function is active.

3. If the function isn't active, set it to active through the boolean and set it back at the end of your function, or else you stop the script.

It's important that there is some form of pause within the function between the setting and reversion of the boolean value; otherwise, its value is immediately changed back. In the following script, we used wait(), which halts the script for approximately one second:

```
local detector = script.Parent
local touchedDebounce = false
local function partTouched(part)
```

```
    if not touchedDebounce then
        touchedDebounce = true
        print("Touched: " .. part.Name)
        wait(1)
        print("Complete")
        touchedDebounce = false
    end
end
detector.Touched:Connect(partTouched)
```

Output with debounce:

```
Touched: Part
Complete
```

▼ TRY IT YOURSELF

Making a Trap Door with .Touched

Let's incorporate what we've learned with .Touched, .CanCollide, and debounce to make a work-ing trap door that drops the player into the unknown. When you're done, feel free to experiment and add to this example. Note that this exercise doesn't necessarily require a debounce, but it's a good example to develop your skills. Keep in mind that these scripts work when anything touches it, including the baseplate. Without collision, the part will fall through.

1. Set up your variables for the script. These include the trap door, your debounce boolean, and the time you want the trap to last. But before doing any of that, make sure you have created a script within the trap:

```
local ACTIVATED_TIME = 1.5
local touchedDebounce = false
local trapDoor = script.Parent
```

2. Connect your .Touched event to your trapActivated() function so you know when someone has touched the trap:

```
local function trapActivated()
end
trapDoor.Touched:Connect(trapActivated)
```

3. Set CanCollide to false and the Traps Transparency to 1 (0-1) for visuals. Then wait the designated time and reset:

```
local function trapActivated()
        trapDoor.Transparency = 1
        trapDoor.CanCollide = false
        wait(ACTIVATED_TIME)
        trapDoor.CanCollide = true
        trapDoor.Transparency = 0
        touchedDebounce = false
```

```
            end
trapDoor.Touched:Connect(trapActivated)
```

4. Add in your debounce so that the trap has time to finish its action before allowing others to fall through. Just like you did before, you check whether debounce is true. If it's not, then set it to true, run your action, and set it back to false.

```
local ACTIVATED_TIME = 1.5

local touchedDebounce = false
local trapDoor = script.Parent

local function trapActivated()
        if not touchedDebounce then
                touchedDebounce = true
                trapDoor.Transparency = 1
                trapDoor.CanCollide = false
                wait(ACTIVATED_TIME)
                trapDoor.CanCollide = true
                trapDoor.Transparency = 0
                touchedDebounce = false
        end
end
trapDoor.Touched:Connect(trapActivated)
```

Try out the script yourself, and you should see it drop the player through the part, as shown in Figure 12.6. Make sure the part is anchored.

FIGURE 12.6
Trap door.

Next, we'll move on to Humanoids and how modifying Humanoid properties makes a more immersive experience—for instance, in the preceding example, you could decrease the player's health when they fall in the trap or make the player tumble.

Introduction to Humanoids

Humanoids are special objects found in player characters and NPCs (non-player characters)—essentially character controllers. They give character models the functionality of characters as you know them with two standard types: R6 and R15. Usually you don't need to worry about Humanoids and characters because they spawn in automatically. But what if you want to modify or use Humanoids outside of the standard rigs to create a more unique immersive experience?

Humanoid Within the Hierarchy

The Humanoid has a couple base expectations to function, which are especially important to pay attention to if you are using custom characters:

▶ The Humanoid expects to be within a model with the PrimaryPart set to the HumanoidRootPart. This is the root driving part of the character that controls the Humanoid's movement in the game. This is generally invisible and placed around the Torso.

▶ Your Humanoid also expects a BasePart named Head to be connected to your Torso/UpperTorso depending on the rig type. If you delete the Humanoid property in the game, you lose all control of the character until Roblox realizes and respawns you.

You'll find the Humanoid object in every character. If you playtest the game and look in Workspace > Your Character, you'll be able to see this property (Figure 12.7).

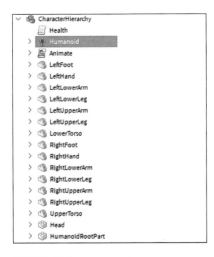

FIGURE 12.7
Character hierarchy.

Humanoid Properties, Functions, and Events

Let's dive into the properties and functions of Humanoid. Within Humanoid, you have many customization options, both directly with its properties and also from the functions and events connected to it. In this section, we get into some of the most commonly used and important parts of Humanoids so you can better understand how to use them within your game. These cover a wide array of useful tools such as applying damage, changing display names, manipulating the camera offset, or reading the Humanoid's state (for example, climbing, dead). Refer to Appendix B for tables of Humanoid properties, functions, and events.

Creating Realistic Walking on Various Surfaces with Humanoid and a Custom Character

You're going to set up a responsive, realistic walking system by modifying the Humanoid and coupling it with a custom character. This should slow down the player on sand and sink the player's feet while reverting to normal on other surfaces. It should also be scalable to allow for easy additions in the future.

1. Using the terrain skills you developed in Hour 5, make a small patch of land using Grass and Sand. Feel free to design it however you want. It doesn't have to be complicated; it should look something like Figure 12.8. Just make sure it has a patch of sand so you can test the script.

FIGURE 12.8
Terrain.

2. Add a script into StarterCharacterScripts so that it appears within your character model when it spawns in. Name the script; in this example, we used SandWalking (Figure 12.9).

FIGURE 12.9
StarterPlayer hierarchy.

3. Set up all the variables that you're going to reference for easy access:

```
local humanoid = script.Parent:WaitForChild("Humanoid")
local hipHeight = humanoid.HipHeight
local walkSpeed = humanoid.WalkSpeed
```

4. Detect changes to Humanoid.FloorMaterial. It's inefficient to run this in a loop, constantly checking, so instead you use GetPropertyChangedSignal, which is fired when the property value changes, and hook it up to a function:

```
function floorMaterialChanged()

end
humanoid:GetPropertyChangedSignal("FloorMaterial"):Connect(floorMaterialChanged)
```

5. Modify a property that is data type and enum.

NOTE

Data Types and Enums

Data types are the different types of data that a variable can store. You can review a list of the Primitive Lua Data Types and Roblox Lua Data types in Appendix A.

Enumerations, or enums, are special data types that store (userdata), a set of values specific to that enum. These are read-only values. To access enums in scripts, we need to use a global object called Enum. You can find the list of enums on the Roblox Developer website here: https://developer.roblox.com/en-us/api-reference/enum.

Check what the floor type is and react accordingly. *If statements* are perfect for this. Because Humanoid.FloorMaterial isn't a string, check it against the Enum Material type:

```
local humanoid = script.Parent:WaitForChild("Humanoid")
local hipHeight = humanoid.HipHeight
local walkSpeed = humanoid.WalkSpeed

function floorMaterialChanged()
        local newMaterial = humanoid.FloorMaterial
        if newMaterial == Enum.Material.Sand then
                humanoid.HipHeight = hipHeight - 0.5
                humanoid.WalkSpeed = walkSpeed - 5
        else
```

```
            humanoid.HipHeight = hipHeight
            humanoid.WalkSpeed = walkSpeed
        end
    end
humanoid:GetPropertyChangedSignal("FloorMaterial"):Connect(floorMaterialChanged)
```

We're taking the current HipHeight and WalkSpeed and reducing it by 0.5 and 5, respectively, while grabbing the saved values and assigning it back for any other floor type.

Great, you're done! Test it out and feel free to add upon it. To add material types, use Enum. Material.MaterialName.

This technique is a really powerful way to improve the player's experience and heighten their immersion with very little effort. This makes them feel like the world is interacting with them on a deeper level. They don't just walk on sand and grass, but they sink in sand and speed up on grass. Although the player may not notice this consciously, these subtle improvements make a big difference to the experience.

But no game is complete without a goal, whether that be to become the richest player in the game, reach the highest level, or complete the story arc. As such, in the following exercise, we show you how to set up a leaderboard and apply a score to said board when a player touches a button.

TRY IT YOURSELF

Keeping Score

Now it's time to combine all the things you've learned to make a score system that awards points every time the player stands on a button. Roblox has a default leaderboard system you can implement, which adds points next to the player's name. As you set up the leaderboard, you want it to run from ServerScriptService, as it's good practice to run from there and the most secure area to store scripts.

1. Set up your part and scripts within the hierarchy so you can edit them later. These scripts and part are the following:

   ▶ One Input, the block/part used as your button (Figure 12.10); Place: Within workspace, Name: PointGiver

   ▶ One script to detect inputs and assign points toward players' scores; Place: Within your button, Name: GivePoint

   ▶ One script to set up the leaderboard to hold players' scores; Place: Within ServerScriptService, Name: SetupLeaderstats

FIGURE 12.10
Button.

2. Roblox picks up additional data to add to the leaderboard through a leaderstats folder (Figure 12.11). Here it collects all ObjectValues, such as StringValue and IntValues. It places the name of the ObjectValue as the header and its value below.

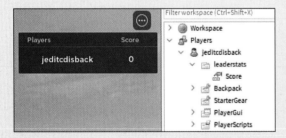

FIGURE 12.11
Leaderboard and player hierarchy with leaderstats.

To set up this system, we need to go back to the script within ServerScriptService to create our leaderstats.

▶ Detect when new players join the game to add the leaderstats folder. To do this, utilize .PlayerJoined as an event of the Players service.

```
local Players = game:GetService("Players")

local function setupLeaderstats(player)
```

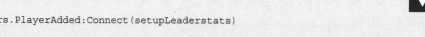

```
        end
        Players.PlayerAdded:Connect(setupLeaderstats)
```

▶ Use instance.new(), which is used to create objects via code—in this case, a folder and an IntValue that will contain the score. When you use instance.new(), you must assign it to a local variable so you can reference it in later lines of code to change its properties and place in hierarchy, like so:

```
local Players = game:GetService("Players")

local function setupLeaderstats(player)
        local leaderstats = Instance.new("Folder")
        leaderstats.Name = "leaderstats"

        local score = Instance.new("IntValue")
        score.Name = "Score"
        score.Parent = leaderstats

        leaderstats.Parent = player
end
Players.PlayerAdded:Connect(setupLeaderstats)
```

3. Go back to the script you made within your button and set up your variables:

   ▶ **Players:** PlayersService, so you can use GetPlayerFromCharacter()

   ▶ **DEBOUNCE_TIME, COOLDOWN_COLOR, POINT_AMOUNT:** Configuration variables

   ▶ **pointGiver, activeColor:** To detect .Touched events and change the part's color

   ▶ **debounce:** To check whether the parts have been touched within the DEBOUNCE_TIME

4. Design the functions and how they interact. You need a function for detecting when your button has been touched, but you also need a function to assign the score to the players leaderstats. To be able to do this, you need a function to find the player within the game.Players service.

```
local Players = game:GetService("Players")

local DEBOUNCE_TIME = 3
local COOLDOWN_COLOR = Color3.fromRGB(255,78,78)
local POINT_AMOUNT = 1

local pointGiver = script.Parent
local activeColor = pointGiver.Color
local debounce = false

local function giveScore(player, POINT_AMOUNT)

end
```

```
local function getPlayerFromPart(part)

end

local function giverTouched(otherPart)

end
pointGiver.Touched:Connect(giverTouched)
```

NOTE
Variable Styling

While not compulsory, Roblox supplies a styling guide for scripters on the platform to use so the code remains consistent across the platform; this covers variables to functions and comments. In doing so, it can save time and provide consistency.

▶ **PascalCase:** Used for enums like objects and classes—for example, Roblox Services

▶ **camelCase:** Used for nonconstant local variables, functions, and member values such as objects

▶ **LOUD_SNAKE_CASE:** Used *only* for local constant variables, such as variables that never change during run time

5. Set up giverTouched(otherPart):

```
local function giverTouched(otherPart)
        if debounce then
                return
        end

        local player = getPlayerFromPart(otherPart)

        if player then
                debounce = true
                giveScore(player, POINT_AMOUNT)

                pointGiver.Color = COOLDOWN_COLOR
                wait(DEBOUNCE_TIME)
                pointGiver.Color = activeColor

                debounce = false
        end
end

pointGiver.Touched:Connect(giverTouched)
```

This function controls detecting inputs, calling other functions, and handling debounce and visual representation.

1. Check whether debounce is activated because if it is, you don't want to continue to run the script. This verification can be done through a check coupled with a return statement that will terminate the script until it's called again.

2. Call the function for fetching the player object from the character, but it may return nil if the part is not connected to a character, which you'll check for in the next line.

3. When you know the player exists and debounce is false, you can start the debounce process. Set it to true and add some points to the player's score using the giveScore() function. Also change the part color as a visual cue to the player that their input was taken, and the part is temporarily deactivated while waiting during the debounce time and resetting everything.

6. Set up getPlayerFromPart(part):

```
local function getPlayerFromPart(part)
        local character = part.Parent
        if character then
                return Players:GetPlayerFromCharacter(character)
        end
end
```

This function controls fetching the player through the character model in workspace, which can be done using the :GetPlayerFromCharacter() function from the PlayersService. The character is made up of an assembly of parts that can be individually detected by .Touched, so if a character touches the button, you can assume it's the part parent— that is, character = part.Parent. In a case where the part is not a part of a character, the PlayersService function returns nil instead, which will be checked for in giverTouched().

7. Set up giveScore(player, POINT_AMOUNT):

```
local function giveScore(player, POINT_AMOUNT)
        local leaderstats = player:FindFirstChild("leaderstats")
        if leaderstats then
                local score = leaderstats.Score
                score.Value = score.Value + POINT_AMOUNT
        end
end
```

8. Now that you know the player, setting the leaderstats score value is fairly simple. First check that leaderstats exist, create a local variable for score, and then add the specified number of points to their score.

Summary

In this hour, we reintroduced collisions and went into detail about how they are handled with more complex objects, such as meshes, unions, and groups. We used what we learned with .Touched, .CanCollide, and debounce to create a working trap door. We also introduced Humanoids and created a realistic walking Humanoid. Finally, we combined everything we learned to make a score system that awards points every time a player stands on a button.

Q&A

Q. What collision fidelity is most costly to performance?

A. PreciseConvexDecomposition

Q. How do you disable an object's collision?

A. CanCollide property

Q. What are the drawbacks of using .Touched()?

A. Unreliable, incorrect detections.

Q. What rig types do Humanoids support?

A. R6 & R15 (/Rthro)

Q. What is the function of a debounce?

A. Limits the rate at which a function can be invoked/executed.

Workshop

Now that you have finished, review what you've learned. Take a moment to answer the following questions.

Quiz

1. How do you show collision fidelity?

    A. Select Show Collisions in Settings.

    B. Select Show Decomposition Geometry in Settings.

    C. Select CanCollide in the object's property window.

2. How should you add collision groups?

    A. Set CollisionGroupId manually.

    B. Use the Collison Groups Editor.

3. Give two alternatives to .Touched().

4. Explain in your own words how loops work.

5. In your own words, what are Humanoids?

6. What is the function of MoveTo() and what two properties does it set?

Answers

1. B. Select Show Decomposition Geometry in Settings.

2. B. Use the Collison Groups Editor.

3. Two alternatives to .Touched() are Magnitude and Region3.

4. Loops are an instruction that is iterated a specified number of times.

5. Humanoids are character controllers.

6. MoveTo() attempts to walk the player character toward a certain position. It sets Humanoid. WalkToPoint and HumanoidWalkToPart.

Exercises

Your task is to create a speed power-up button that deactivates for two seconds after last use.

1. Place down two or more parts to act as your buttons, placing within a folder.

2. Add a script within the folder, and set up a debounce variable (boolean).

3. Use a for loop on the folder, validating the objects are BaseParts.

4. Within it, set up .Touched events.

5. Make sure the .Touched part is a character part and Enabled is true.

6. Set Enabled to false, change the Humanoid property, and use wait().

7. Set Enabled to true.

In the next exercise, create a door that only unlocks when a certain player touches it and kills other players.

1. Place down two or more parts to act as your doors, placing within a folder.

2. Add a script within the folder, naming it appropriately.

3. Use a for loop on the folder, validating the objects are BaseParts.

4. Within it, set up the .Touched event and set up a debounce variable (boolean).

5. Make sure the .Touched part is a character part.

6. Verify the character is the correct player.

7. Set debounce to true, and if the player is the correct one, then temporarily set CanCollide to false.

8. Otherwise, remove 20 health from the character using TakeDamage().

HOUR 13
Interacting with GUIs

What You'll Learn in This Hour:

▶ How to create GUIs
▶ What the basic GUI elements are
▶ How to code interactive GUIs
▶ How to tween
▶ What layouts and constraints are
▶ How to make a GUI countdown

So far you've learned how to build things in your game and add terrain to create an environment. You've also learned how to code things to add functionality and interactivity, but there's one main thing missing—a Graphical User Interface or GUI, which is used to display images and text on the player's screen. In this hour, you find out how to create UIs, how to code interactive GUIs, and how to add layouts/constraints. Adding a GUI to your game is essential for tutorials, displaying information, or selling items in a shop. Figures 13.1 and 13.2 show a couple of examples of GUIs.

FIGURE 13.1
GUI example from *Shoot Out!* The left side shows health, and the right side shows controls for the game and ammo.

FIGURE 13.2
This GUI example from *Build It, Play It: Island of Move* by Roblox Resources includes the E floating above the NPC's head and the Click to Interact button.

Creating GUIs

There are three types of GUIs in Roblox, all of which work in very similar ways:

- **SurfaceGui** is where a GUI is displayed on a surface in the 3D environment.

- **ScreenGui** is where a GUI is displayed on the screen as a 2D UI.

- **BillboardGui** is essentially a 3D GUI that hovers above a part that you *adorn* it to.

PlayerGui

Let's start off by creating a 2D PlayerGui. These are typically used to give players information about their characters, such as score, health, or gold, as shown in Figure 13.3. Since we want this to be displayed as a 2D element on the screen, we create it using ScreenGui objects.

FIGURE 13.3
PlayerGui example from *Digital Civility Scavenger Hunt* by Roblox Resources: The purple icons at the top that provide player information are made with ScreenGui objects.

Use the following steps to create a ScreenGui:

1. In Explorer, insert a ScreenGui into StarterGui (Figure 13.4).

FIGURE 13.4
Insert a ScreenGui.

Anything inside this ScreenGui is displayed to the player so long as Enabled is checked, as in Figure 13.5. If the option is unchecked, players don't see anything parented to the ScreenGui.

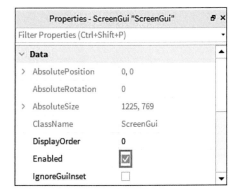

FIGURE 13.5
Enable or Disable a ScreenGui to hide or show all its children.

2. Insert a TextLabel into this ScreenGui (Figure 13.6) to display text to the player. Try adjusting the Font and TextColor3 properties. You can also adjust the TextTransparency and add a border by adjusting the BorderSizePixel and BorderColor3: If you have a lot of text, enable TextWrapped.

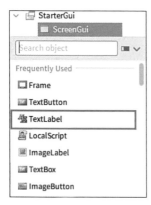

FIGURE 13.6
Insert a TextLabel into the ScreenGui.

3. Also adjust the size and position of the TextLabel. To do this, you can change the size and position properties in the property window (Figure 13.7). Both position and size use UDim2 values, which is a format that looks like {0,0}, {0,0}. The first number is scale and the second is offset. Scale works using any screen size and calculating the percentage of the screen size. For example, if you set the size of the frame to {0.5,0}, {0.5}, it will be 50% the height and 50% the width of the current screen size. Scale is especially useful for designing games that work on multiple platforms.

FIGURE 13.7
Adjust Size and Position.

4. Alternatively, you can resize Offset by setting the Size or Position to a specific number of pixels (Figure 13.8). This doesn't work so well on varying screen sizes; however, it can be useful in some instances—for example, creating a 50-pixel margin around a frame regardless of the screen size.

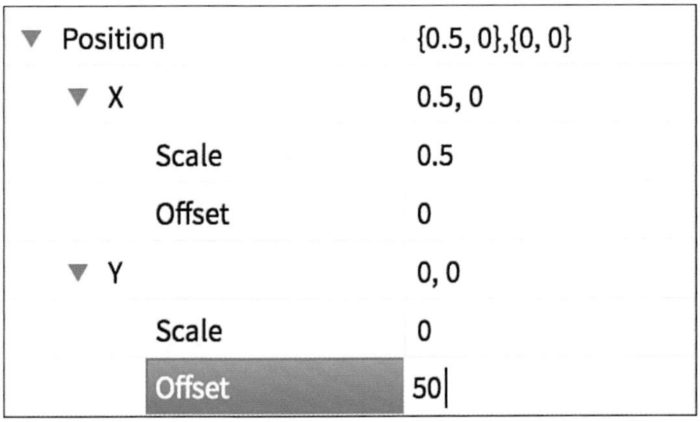

▼ Position	{0.5, 0},{0, 0}
▼ X	0.5, 0
Scale	0.5
Offset	0
▼ Y	0, 0
Scale	0
Offset	50|

FIGURE 13.8
Adjust Offset.

5. Adjust AnchorPoint (Figure 13.9) to set the center point of a GUI, such as for hanging a clock or picture frame on the wall. For example, you can use an AnchorPoint of {0.5, 0.5} to align the element to the top center of the screen. Assuming Position is also set to {0.5, 0, 0.5, 0}, the element will be in the middle of the screen.

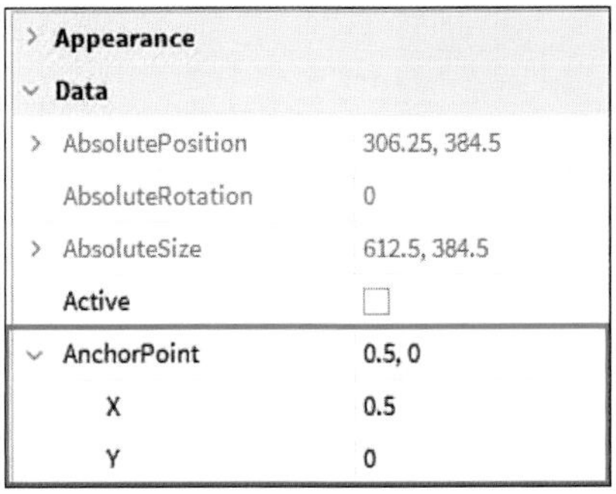

> Appearance	
˅ Data	
> AbsolutePosition	306.25, 384.5
AbsoluteRotation	0
> AbsoluteSize	612.5, 384.5
Active	☐
˅ AnchorPoint	0.5, 0
X	0.5
Y	0

FIGURE 13.9
Adjust AnchorPoint.

6. If you have multiple images stacked on top of each other, the order might seem random. To adjust this, you can change the ZIndex property: Any GUI element with a lower ZIndex value appears under another element with a higher ZIndex value (Figure 13.10).

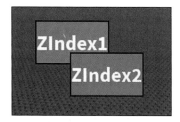

FIGURE 13.10
Example showing the impact of using ZIndex on two TextLabels.

SurfaceGui

You also can use a SurfaceGui to display text and images to a player. However, instead of displaying 2D on a screen, it is displayed on a surface in the 3D environment, as in Figure 13.11.

FIGURE 13.11
Digital Civility Scavenger Hunt by Roblox Resources: 2D GUI displayed on a part as a sign.

There are two ways to do this depending on whether the sign will be static or the player will be able to interact with it. For static billboards or signs that don't require interaction, you parent the SurfaceGui to the part to which you want the SurfaceGui to be displayed.

To parent the SurfaceGui to its part, begin by placing the SurfaceGui:

1. Insert a SurfaceGui into a part.

2. In the SurfaceGui, insert a TextLabel (Figure 13.12).

FIGURE 13.12
TextLabel inserted in the SurfaceGui.

3. Change which side of the part the label shows up on by selecting the SurfaceGui and changing the Face property (Figure 13.13).

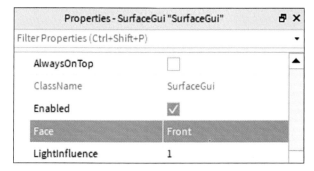

FIGURE 13.13
Changing the face property of the SurfaceGui.

Once the SurfaceGui is successfully parented, you need to size the TextLabel to take up the entire side of the part. To do this, you need to know two pieces of information: the size of your part and how many pixels per stud the label is. The first you have to find out for yourself. However, unless you change it, the label will use 50 pixels per stud. Follow these steps:

1. In the part's properties, note the size of the relevant axis. For the part shown in Figure 13.14, X equals 9 studs, and Y equals 4 studs. Remember, your face might be on a different axis.

2. Select the TextLabel. In Properties, scroll to Size.

3. For both the X offset and Y offset, use NumberOfStuds x 50. For the example shown in Figure 13.15, that means that X equals 450 (9 studs × 50), and Y equals 200 (4 studs × 50).

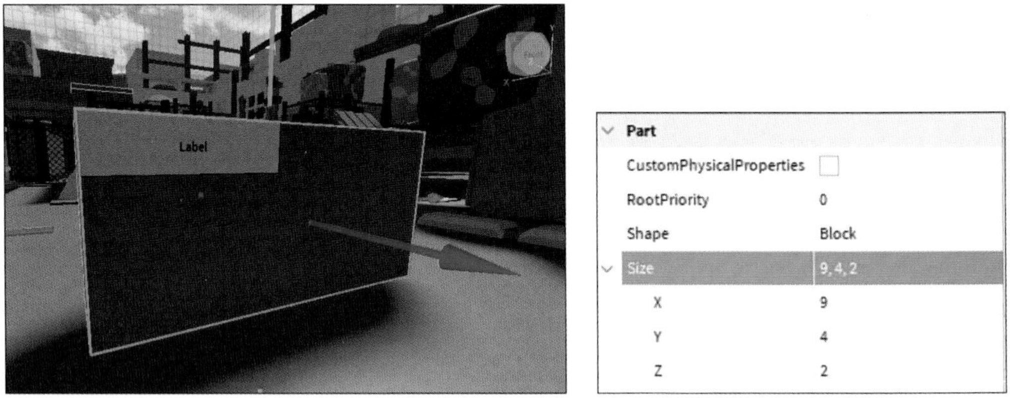

FIGURE 13.14
Sizing the TextLabel for the SurfaceGui.

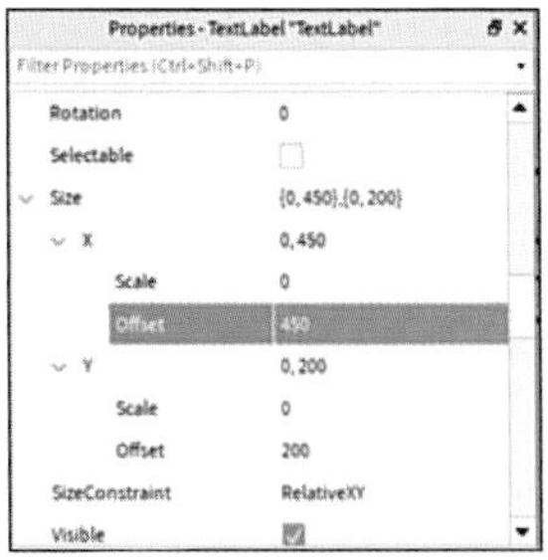

FIGURE 13.15
Use Offset to size the TextLabel.

When the player interacts with the GUI, such as pressing a button or when the text has to update during the game, there is a different process. For an interactive GUI, you need to keep the SurfaceGui in StarterGui but set the Adornee to the part (Figure 13.16). The Adornee value is the part that the SurfaceGui should be displayed on.

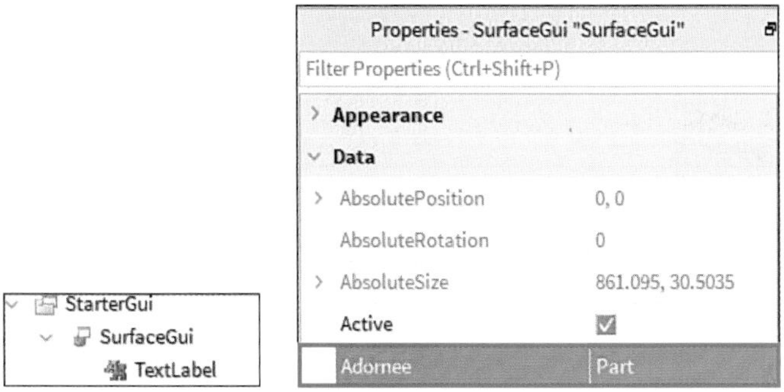

FIGURE 13.16
Setting the Adornee of a SurfaceGui.

Another interesting thing you can do with a SurfaceGui is adjust the LightInfluence value (Figure 13.17), which helps create a bright billboard effect by adjusting how much light influences the SurfaceGui.

Enabled	☑
Face	Front
LightInfluence	1

FIGURE 13.17
Adjusting the LightInfluence.

For example, if you put a PointLight inside the part with the SurfaceGui and change LightInfluence to 0.1, the SurfaceGui remains visible no matter how dark the environment is. Figure 13.18 shows a comparison of how the LightInfluence can affect the brightness effect.

FIGURE 13.18
Comparison of different LightInfluence values.

Basic GUI Elements

Every GUI is made up of certain elements such as Frames and TextLabels. The items in the following list are the different basic elements that you can put together to create complex but beautiful user interfaces in your game.

▶ A TextLabel is used to display text to the player.

▶ A TextButton is the same except a player can hover over and click it. Later in this hour, we explain how you can trigger an event when a player clicks the button.

▶ You use an ImageLabel to display an image.

▶ An ImageButton is the same as an ImageLabel, but a player can hover and click it.

▶ Frames are really useful because they can hold multiple labels or buttons inside them. You are also able to use Layouts, which we will look at in a bit.

Coding Interactive GUIs

GUIs can't always be static images that don't do anything; they often need to make something happen. In this case, you create a button like the ones in Figure 13.19 that will open and close an imaginary shop menu.

FIGURE 13.19
Build A Boat for Treasure: Clickable GUI buttons to open the shop and other menus.

To represent the shop menu, you use a Frame object. These are great for organizing multiple GUI elements, such as different items to sell. Then you make a button that opens the shop when a user clicks it and closes the shop when the user clicks the button again.

First, set up the ShopGui by following these steps:

1. Insert a ScreenGui into StarterGui. Rename it ShopGUI.

2. Select the ScreenGui, and add an ImageButton. Rename it ShopButton (Figure 13.20).

FIGURE 13.20
Insert an ImageButton and rename it.

3. Add a frame to act as the shop (Figure 13.21). You will make the frame disappear and reappear when the player clicks the button. Make sure that the frame isn't right on top of the ImageButton so you can see both at the same time.

FIGURE 13.21
Insert a frame into the ShopGui.

4. Select the frame. In Properties, scroll-down and uncheck the Visible option (Figure 13.22). This way, the player won't see the shop until they click the button.

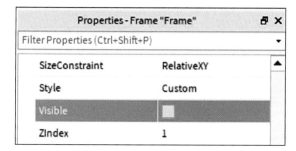

FIGURE 13.22
Make the shop invisible until clicked.

5. Add code that will register when a player has clicked or tapped the button to make the shop visible to the player. To begin, insert a LocalScript into the ShopButton you created, as shown in Figure 13.23.

FIGURE 13.23
Insert a LocalScript into the button.

6. Use the Activated event to register when the player has clicked or pressed the button. Inside that event function, you can change the visibility of the frame when clicked. This particular example sets the visibility to the opposite of the current visibility. Type the following code:

```
local ImageButton = script.Parent
local ScreenGui = ImageButton.Parent
local Frame = ScreenGui.ShopFrame

local function buttonActivated()
      Frame.Visible = not Frame.Visible
end

ImageButton.Activated:Connect(buttonActivated)
```

7. Also make the button change color slightly when the player hovers over it. To do that, use the MouseEnter and MouseLeave events to detect when the player hovers the mouse or stops hovering the mouse over the button:

```
local ImageButton = script.Parent
local ScreenGui = ImageButton.Parent
local Frame = ScreenGui.ShopFrame

local function buttonActivated()
      Frame.Visible = not Frame.Visible
end

local function mouseEnter()
      ImageButton.ImageColor3 = Color3.fromRGB(25, 175, 25)
end

local function mouseLeave()
      ImageButton.ImageColor3 = Color3.fromRGB(255, 255, 255)
end

ImageButton.Activated:Connect(buttonActivated)
ImageButton.MouseEnter:Connect(mouseEnter)
ImageButton.MouseLeave:Connect(mouseLeave)
```

8. Start a playtest. You should be able to click the image button to make the shop frame appear, and click it again to make it go away.

Tweening

Another really useful thing you can do with GUIs is something called tweening, which is where you animate, or move, a GUI element. For example, you could make a GUI slide onto the screen and then bounce. You do this with TweenPosition and TweenSize, or do both using TweenSizeAndPosition.

Add the highlighted lines of the following code to your script to make the button larger when a player is hovering over it:

```
local function mouseEnter()
    ImageButton.ImageColor3 = Color3.fromRGB(25, 175, 25)
    ImageButton:TweenSize(UDim2.new(0,110, 0, 110), nil, nil,.25) -- Active size
end

local function mouseLeave()
    ImageButton.ImageColor3 = Color3.fromRGB(255, 255, 255)
    ImageButton:TweenSize(UDim2.new(0, 100, 0, 100),nil, nil, .25) -- Original
        size

end
```

Above, UDim2 is used to hold the X and Y values for Scale and Offset. You can use the Property window (Figure 13.24) to find the numbers you want to use; just scale the GUI to the desired size and copy the four values in the order you see them.

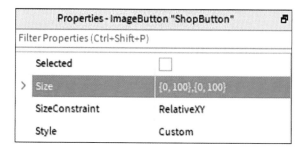

FIGURE 13.24
Scale the GUI to the desired size.

Playtest to see how you like the results. To make the animation slower or faster, modify the last number (.25). This is the amount of time in seconds it takes to complete the tween.

To learn different ways you can make GUIs react and slide or bounce onto screen, look up "Tween easing styles" on the Roblox Developer Hub.

Try Different Tweens, Including TweenSize and TweenSizeAndPosition
Use different easing styles and try to implement a callback function that does something interesting when the tween has finished!

Layouts

Roblox offers different layouts that you can use with GUIs. Layouts are extremely useful because you don't have to spend hours writing a script to automatically resize or position elements based on the size or number of them. Let's go through the layouts you can use:

▶ **UIGridLayout:** Especially useful in a Frame or ScrollingFrame. It organizes all elements into a grid shape. You can set the CellSize and Padding to determine how large each element is and how close together they are. Other options include SortOrder, which changes the order depending on the name of the element, and LayoutOrder, which you can use to control the order in which elements are displayed. An example of UIGridLayout is shown in Figure 13.25.

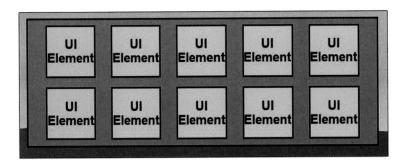

FIGURE 13.25
An example of the UIGridLayout in use.

▶ **UIListLayout:** Arranges elements inside a Frame or ScrollingFrame into a list, which is especially useful for ScrollingFrames because you can add as many elements as you want, and they all line up nicely in a list. Again, you can adjust the Padding for each element along with VerticalAlignment and HorizontalAlignment to determine where the list will align with. An example of UIListLayout is shown in Figure 13.26.

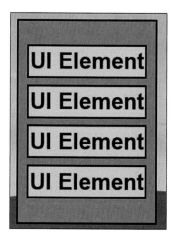

FIGURE 13.26
An example of the UIListLayout being used.

▶ **UITableLayout:** Similar to the grid layout in that it lays out UI elements in a grid shape. The UITableLayout arranges elements into rows and then arranges the children of those elements into columns. Figure 13.27 shows an example of UITableLayout, and Figure 13.28 shows how UITableLayout works.

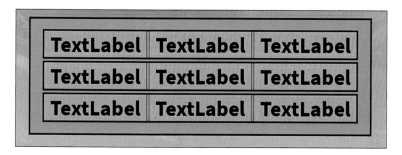

FIGURE 13.27
An example of the UITableLayout being used.

▶ **UIPageLayout:** Organizes elements into a carousel of elements the user can scroll through. The main benefit of this is that it works really well with mobile devices and controllers for console devices. Figure 13.29 shows an example.

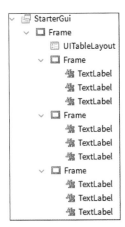

FIGURE 13.28
Showing how the UITableLayout works.

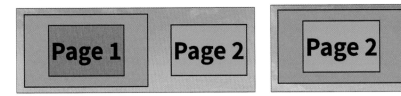

FIGURE 13.29
An example of the UIPageLayout being used.

Roblox also has some really useful constraints that you can use to make GUIs. These constraints generally keep certain values (such as size and position) between certain specified levels:

▶ **UIAspectRatioConstraint:** When you insert this constraint into an element, such as a frame, it keeps that element in the aspect ratio you set by adjusting the size, regardless of the screen size. (See Figure 13.30.)

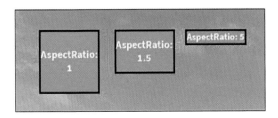

FIGURE 13.30
Using the UIAspectRatioConstraint.

▶ **UITextSizeConstraint:** Inserting this constraint into a TextLabel or TextButton keeps the font size between the MaxTextSize and MinTextSize you set. This can be useful for preventing font sizes from becoming unreadable on, for example, large TV screens that console players might use. The general rule of thumb with this constraint is to make sure the text is legible from about 10 feet away. Compare the two examples of UITextSizeConstraint in Figure 13.31.

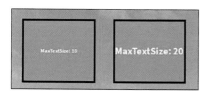

FIGURE 13.31
Comparison of two TextLabels using different UITextSizeConstraint values.

▶ **UISizeConstraint:** This constraint works similarly to the text size constraint, but it keeps the UI element size, rather than the text, between MaxSize and MinSize. This constraint works with absolute pixel sizes, so if you set the MaxSize to {50, 70}, the size of the UI will not be wider than 50 pixels or taller than 70 pixels. See the examples in Figure 13.32.

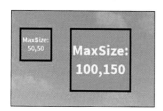

FIGURE 13.32
Using a UISizeConstraint.

Making a GUI Countdown

Now that you've learned about what GUIs are, how to make them, and how to code them, you can make a countdown to display on the player's screen. This could be used in a variety of games, such as a round-based game that ends when the timer reaches zero. For now, you should try making this as a PlayerGui, but you could also try a SurfaceGui.

1. Insert a ScreenGui and rename it to something suitable, such as Timer or Countdown (Figure 13.33).

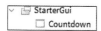

FIGURE 13.33
Insert a ScreenGui and rename it.

2. Insert a frame into the ScreenGui (Figure 13.34) to act as the background for the timer. You can adjust the size using Scale and change the color using BackgroundColor3.

FIGURE 13.34
Insert a frame into the ScreenGui.

3. Also inside the frame, add a TextLabel (Figure 13.35) to actually display the countdown. Again, adjust the size, position, color, and font of this until you're happy with how it looks.

FIGURE 13.35
Add a TextLabel into the frame.

4. Add a LocalScript into the TextLabel so you can code the functionality of the timer (Figure 13.36).

FIGURE 13.36
Add a LocalScript into the TextLabel.

Now comes the coding part. You should use a loop to create the timer. Use a while loop with a wait(1) and decrease a variable every second. You can then set script.Parent.Text equal to that variable. Some basic example code is shown here:

```
local textLabel = script.Parent
for timer = 60, 0, -1 do
    wait(1)
    textLabel.Text = Timer
end
```

Summary

In this hour, you've learned how to add GUIs to your game. You've been introduced to the different types of GUI elements and how to use them in both Surface and Player GUIs to display images and text. You've also looked at different layouts and constraints that can be used to arrange your user interface and scale it correctly on different screen sizes. Finally, you've learned how to code GUIs to allow players to interact with them.

Q&A

Q. Should I use Scale or Offset to size my UI?

A. We usually recommend using Scale because it keeps the GUI relatively the same size regardless of screen size. However, you can also use Offset with a UIScale object, which can be adjusted depending on screen size.

Q. How much memory do ImageLabels and ImageButtons take up?

A. To keep memory usage to a minimum, you should try to upload images in low resolution. You should also keep use of ImageLabels, ImageButtons, Textures, and Decals to a minimum and reuse the same image where possible (for example, using one UI background image for all menus).

Workshop

Now that you have finished this hour, take a few moments to review and see if you can answer the following questions.

Quiz

1. What is the name of the Roblox instance that displays text to a player?

2. GUI stands for _____ user interface.

3. What type of script should you use for coding GUI interaction?

4. True or False: A SurfaceGui is displayed on a part.

5. A ScreenGui must be in _____ to be displayed to the player.

6. True or False: Offset size/position adjusts the absolute number of pixels.

7. True or False: Any GUI element with a lower ZIndex value appears under another element with a higher ZIndex value.

Answers

1. TextLabel is the name of the Roblox instance that displays text to a player.

2. GUI stands for graphical user interface.

3. You use a LocalScript for coding GUI interaction.

4. True. SurfaceGuis are displayed on parts, and the PlayerGui is displayed as a 2D interface.

5. A ScreenGui must be in a StarterGui to be displayed to the player.

6. True. Offset size/position adjusts the absolute number of pixels.

7. True. Any GUI element with a lower ZIndex value appears under another element with a higher ZIndex value.

Exercises

Build on from the GUI countdown you made earlier to add some extra functionality to the count-down timer. Add a button which, when clicked or tapped, closes the countdown timer frame.

1. Using the countdown you made earlier, insert a TextButton into the ScreenGui (which was named "Timer" or "Countdown").

2. Insert a LocalScript into the TextButton.

3. Use the Activated event to change the Visible property of the frame.

4. You can use the not keyword to change the visibility. For example,

```
local button = script.Parent
local screenGui = button.Parent
    local frame = screenGui.Frame

    frame.Visible = not frame.Visible
```

If Visible is true, it will be set to false, and vice versa.

Bonus Exercise: Create a SurfaceGui with a TextButton that prompts the purchase of a GamePass when clicked.

1. Create a new part and parent it to the workspace.

2. Change the Color3 and Material if you want!

3. Insert a SurfaceGui into the StarterGui and click the empty box next to Adornee. You should then click the part you created to select it, and the Adornee value will be set to that part.

4. Adjust the Face value of the SurfaceGui to make sure it is displaying on the correct side of the part.

5. Add a TextButton into the SurfaceGui and adjust the styling however you like.

6. Insert a LocalScript into the TextButton.

7. Using the Activated event and the MarketPlaceService and the PromptGamePassPurchase functions, prompt the purchase of a GamePass when clicked.

NOTE

Prompting the GamePass Purchase

By putting the SurfaceGui in the StarterGui and setting the Adornee value to the part, this allows you to use a LocalScript to locally prompt the GamePass purchase.

HOUR 14
Coding Animation

What You'll Learn in This Hour:

▶ How to work with CFrame position

▶ How to work with CFrame rotation

▶ How to tween parts

▶ How to use SetPrimaryPartCFrame()

Animations can be an extremely useful tool in your games, from bringing characters to life to making small visual feedback loops for the player. A video game is best defined as an "interactive experience," and you need to allow the player to interact with the world in as many ways as possible by provoking whatever senses are at your disposal. As with films, the main senses that developers can trigger in games are visual and auditory. As such, you can use animation—and other visual and auditory cues—to signal emotions, actions, and more.

There are many cases where animating by hand is the best method of achieving this effect, such as with character expressions and predetermined movements. This hour explains the basics of coding animations.

Working with Position and Rotation

One of the best ways to make your world come alive is to animate the objects within it. Buttons should move when pressed, doors should open, and monsters should take a swing at you. To move objects in a game, there are two main properties that need to be affected by code: Position and Rotation. As you can see in Figure 14.1, every part in the game has Position and Orientation values for the X, Y, and Z axes.

Properties - Part "Part"	
Filter Properties (Ctrl+Shift+P)	▼
∨ Orientation	15, 0, 25
X	15
Y	0
Z	25
Parent	ExampleModel
∨ Position	28, 16, 45
X	28
Y	16
Z	45

FIGURE 14.1
Position and Orientation values for parts in a game.

In Figure 14.2, you can see the X axis (red) and Z axis (blue) are side to side, whereas the Y axis (green) runs up and down.

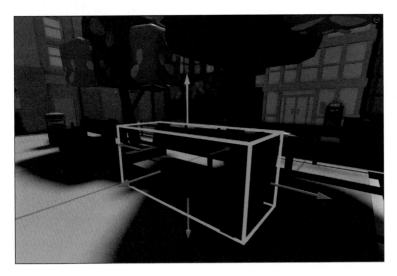

FIGURE 14.2
Moving an object in *Digital Civility Scavenger Hunt* by Roblox Resources.

The information for these values is held in a data type called CFrame, which is short for coordinate frame. One way to change rotational and positional values is to provide a new CFrame with the desired coordinates and rotation. The format for creating new CFrames is `CFrame.New(X,Y,Z)` where X, Y, and Z can be variables or numbers.

Moving an Object from Point A to Point B

If all you want to do is set the position of a part to a specific X, Y, Z location, you just need the following code:

```
part.CFrame = Cframe.new(0, 0, 0) --Replace 0's with X, Y, Z values
```

Quite often, though, you need to move a part a small amount relative to its *current* position. Figure 14.3 shows a big red button (on the left) that, when clicked, moves slightly downward and changes color from red to green (on right).

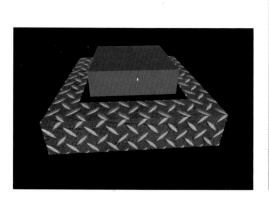 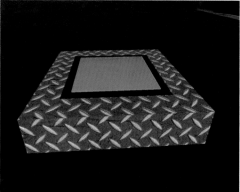

FIGURE 14.3
Changing a button's position and color to demonstrate it has been clicked.

Setting Up the Button Model

To create a button players can interact with, you need separate parts for the button and for the button socket. To make the buttons function, you use a script in combination with a ClickDetector object to detect when a player has clicked or pressed the button. ClickDetectors are platform agnostic, meaning they work whether the player is on a mobile device, PC, or Mac.

1. Create a button and a socket base for it to sink into. Make sure the button is a single part because models are handled slightly differently.

2. To detect whether the player clicks the button, inside the button add a ClickDetector and a script (Figure 14.4).

FIGURE 14.4
A script and a ClickDetector parented to the part to be used as a button.

3. Add the following code. Anything inside onClick() runs when the button is clicked.

```
local button = script.Parent
local clickDetector = button.ClickDetector

local function onClick()
        print("button was clicked")
end

clickDetector.MouseClick:Connect(onClick)
```

NOTE

Use a Part

This script only works when parented directly to a part, not an entire model.

Setting a New CFrame

Now, it's time to get the button's current CFrame and assign a new frame when the player clicks. First, though, you need to determine which direction you want your part to move in. Because the part moves relative to itself, use the local translation/rotation tools. You should see an L in the bottom-right corner of the selected object (Figure 14.5). If you don't, press Ctrl+L or Cmd+L to switch modes.

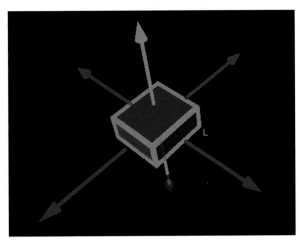

FIGURE 14.5
Active Translate widget showing an L in the bottom-right corner of the object to indicate Local mode.

In the following example, the Y axis is the direction the part should move in:

1. Determine the axis you want your part to move in. Make sure you're using local transform tools.

2. Beneath the `print` statement, add the highlighted line to get the current CFrame and set how much you want to offset the part using `CFrame.new`. For this example, the part will move -0.4 studs along the Y axis.

```
local function onClick()
        print("button was clicked")
        button.CFrame = button.CFrame * CFrame.new(0, -0.4, 0)
end
```

3. Playtest your code. You may need to play with the values to get a button that moves in the right way.

TRY IT YOURSELF ▼

Single Button Clicks

You may notice you can keep clicking and the button keeps moving. See if you can figure out how to change it so the button can only be clicked once.

Rotating Parts with CFrames

The basic formula for rotating parts is

```
local part = script.Parent
part.CFrame = part.CFrame * CFrame.Angles(0,0, math.rad(45))
```

Here, a part is rotated 90 degrees on the Z axis. Again, CFrames are used, but now `CFrame.Angles()` is being used for rotation. This function uses radians instead of degrees for each of the three axes. So that you don't have to worry about how radians work, `math.rad()` converts degrees to radians for you.

▼ TRY IT YOURSELF

Adding Rotation to a Part

When you insert a script into a part and test the rotation, you'll notice the part rotates from its center. Rather than have the part rotate from the center point, imagine a hatch leading to an underground tunnel with a lid (Figure 14.6). The player needs to click the lid to open the hatch. To make the lid work properly, the lid will have to rotate onto its edge.

FIGURE 14.6
An underground tunnel with a lid.

1. Create a simple lid for the hatch using a cylinder (Figure 14.7). Once again, just use a simple part.

FIGURE 14.7
A cylinder for the lid.

2. Insert a ClickDetector and a script into the lid part.

3. Copy the following to make the lid rotate when clicked:

```
local lid = script.Parent
local clickDetector = lid.ClickDetector

local function onClick()
        lid.CFrame = lid.CFrame * CFrame.Angles(0,0, math.rad(90))
end

clickDetector.MouseClick:Connect(onClick)
```

4. Test your code. If you don't notice a change, try rotating a different axis. You can check which axis to use with the local (Ctrl+L/Cmd+L) rotation and move tools.

You may also notice that because the lid rotates from the center, it gets stuck in the ground (Figure 14.8).

FIGURE 14.8
The lid stuck in the ground.

To fix that, you need to add an offset to the point from which the lid rotates by multiplying by an additional CFrame, like so:

```
lid.CFrame * CFrame.Angles(0, 0, math.rad(90)) * CFrame.new( 0, -9, 0)
```

Add an offset to the lid:

1. Figure out which axis the lid should move along. Hint: It probably won't be the same one you're rotating on.

2. Determine how much you need to move the object. This part is 18 studs long, so half would be 9 studs in the Y axis.

3. Add `* CFrame.new( 0, -9, 0)` to the end of the CFrame line (Figure 14.9) using the X, Y, and Z values that work for you.

```
CFrameFile.rbxl ×    HingeScript ×
1    local lid = script.Parent
2    local clickDetector = lid.ClickDetector
3
4  ∨ local function onClick()
5        print("lid was clicked")
6        lid.CFrame = lid.CFrame * CFrame.Angles(0,0,  math.rad(90)) * CFrame.new( 0, -9, 0)
7    end
8
9    clickDetector.MouseClick:Connect(onClick)
10
```

FIGURE 14.9
Image of code with the added CFrame for translating the lid to the correct location.

4. Playtest and test your code. You may find that you need to move the part in more than one axis. The finished product should look similar to Figure 14.10.

FIGURE 14.10
The open lid.

This is just the beginning of CFrames. For more information, go to the Roblox Developer Hub, Understanding CFrames at https://developer.roblox.com/articles/Understanding-Cframe.

Moving Objects Smoothly with Tween

Assigning a new CFrame makes objects jump into place. If you want them to move smoothly, you need to use tweens. This is similar to how tweens were used with GUIs in Hour 13. To recap, tweens transition properties from a starting value to a goal value. Size, Position, Rotation, and Color are all properties that can be tweened. In this section, you find out how to tween between two CFrames.

Setting up a tween looks like Figure 14.11.

```
local tween = TweenService:Create(part, tweenInfo, goal)
```
Variable Identifier   Service   Function   Parameters

FIGURE 14.11
Setting up a tween.

`TweenService:Create` is the sole and extremely powerful function for creating tweens. It has three parameters:

▶ **Part Instance:** The instance to be tweened

▶ **tweenInfo:** Special type, `TweenInfo.new()`

▶ **Goal Dictionary:** A table of properties to be changed and goal values to reach

It's good practice to assign these values to variables that can then be input as each parameter, like so:

```
-- Get the TweenService
local TweenService = game:GetService("TweenService")

-- Part to be tweened
local part = workspace.Part -- Parameter #1

-- How the tween will behave

local tweenInfo = TweenInfo.new(1) -- Parameter #2
local goal = {Position = 1,0,0} -- Parameter #3

-- Create the tween and play it
local Tween = TweenService:Create(part, tweenInfo, goal)
Tween:Play()
```

Tweening Between Two Points

An object can be made to endlessly transition between two points by setting a goal value and taking advantage of the different tween parameters. This example demonstrates a platform that slides back and forth so that players can cross a canyon (Figure 14.12).

FIGURE 14.12
The pink platform can cross the canyon.

To create a moving platform, do the following:

1. Create the platform using a single part.

2. Insert a script into the part. Copy the following code to call the Tween service and get the part to be tweened:

```
local TweenService = game:GetService("TweenService")
local platform = script.Parent
```

3. Tween info actually has six customizable parameters. Set them up as shown here. Adjust the time in seconds as needed, and don't worry about the rest for now.

```
-- How the tween will behave
local tweenInfo = TweenInfo.new(
      10,                       -- Time it takes to reach goal
      Enum.EasingStyle.Linear,  -- Easing style
      Enum.EasingDirection.In,  -- Easing direction
      -1,                       -- Number of times repeated; -1 is infinite
      true,                     -- If true, reverses after goal
      0.5                       -- Delay before tween plays
)
```

4. Anything that you want the tween to modify goes inside a table. Add the following code to your script, and use the desired X, Y, Z values inside `Vector3.new()`:

```
-- Add goal values to a table
local goal = {}
goal.Position = Vector3.new(-191, 35, 39.6)
```

5. Create the tween and have it play.

```
--  Create the tween
local tween = TweenService:Create(platform, tweenInfo, goal)
tween:Play() -- Plays the tween
```

6. Playtest your code and adjust the time in seconds as needed.

TRY IT YOURSELF ▼

Adding the Color Property

Try adding the Color property to the goal table as well.

Easing Style and Direction

By default, the tween will smoothly transition between two values. However, this doesn't match real life where things don't start and stop motion in a perfectly uniform manner. For example, a train moves slowly in the beginning, speeds up, and then slows back down before reaching its final destination. EasingStyle gives you options about how the tween behaves before it reaches its goals, such as slowing down before stopping, speeding up, or even overshooting and snapping back. Figure 14.13 shows how different tweening styles reach their goals over time. The bottom left of the square is the starting value; the top right is the goal value.

EasingDirection controls the direction in which the graph plays. For example, if the tweening style is elastic and you're working with position, In would make the object shake back and forth at the end. Out would play the graph backward so the shaking would happen at the beginning. InOut would shake at both sides.

▼ TRY IT YOURSELF

Playing with Tween Style and Direction

Try out a few different tween style and tween direction combinations on your moving platform.

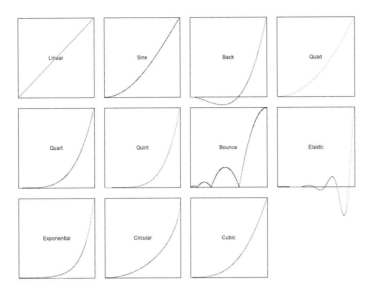

FIGURE 14.13
Transitions for tweens.

Moving an Entire Model

Moving an entire model is a little different than moving a single part. If you look at a model in Explorer, you'll see that it doesn't have Position or Orientation properties. Instead, to move the model, you move the primary part (Figure 14.14). This is one reason setting the primary part and making sure everything is welded is important.

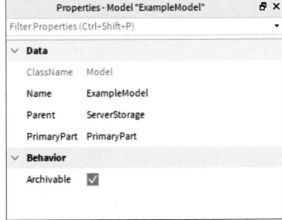

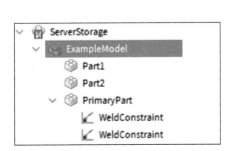

FIGURE 14.14
The properties of a model object.

To move the model, use `SetPrimaryPartCFrame()`. For example,

```
local model = script.Parent
local newCFrame = CFrame.new(0 ,20, 0)

model:SetPrimaryPartCFrame(newCFrame)
```

The following code snippet drops a copy of a model from the sky every second. The model needs to be stored in ServerStorage:

1. Place a model into ServerStorage. Make sure it's not anchored so it will fall freely.

2. In ServerScriptService, add a script.

3. Copy the following code. Modify the name of the model to match yours and set the drop location.

   ```
   local ServerStorage = game:GetService("ServerStorage")
   local modelToDrop = ServerStorage:WaitForChild("ExampleModel")

   while true do
           local newCopy = modelToDrop:Clone()
           newCopy.Parent = workspace
           newCopy:SetPrimaryPartCFrame(CFrame.new(-100, 40.957, -108))

           wait(1)
   end
   ```

4. Playtest.

TIP

Welding and Unanchoring

If your model falls apart, make sure it's welded together properly. If parts stick in the air, make sure they are unanchored.

Summary

In this hour, you've learned to create animations with code that make your game world more interactive. You started by updating a part's CFrame using `CFrame.new()` and then built upon that. You rotated a CFrame for an object using `CFrame.Angles(0, 0, math.rad(90))`, where `math.rad()` takes in the more familiar degree values and converts them to radians.

Finally, you learned how to take into account whether you are working with a model or an individual part. Models do not have position or rotation information of their own. Instead `SetPrimaryPartCFrame()` is used to find and manipulate the primary part of an object.

Q&A

Q. What is the purpose of EasingStyles?

A. To allow tweens to progress to their goal in ways that are more natural than just linearly.

Q. What two components are CFrames made up of?

A. Positional and rotational.

Workshop

Now that you have finished, review what you've learned. Take a moment to answer the following questions.

Quiz

1. Give three examples of properties that can be tweened.

2. What function is used to set up a tween?

3. What do EasingDirections do?

4. What constructor is used to create a blank CFrame?

Answers

1. Size, Transparency, and Position are properties that can be tweened.

2. The `TweenService:Create()` function sets up a tween.

3. EasingDirections control the direction the tween goes through the graph as it plays.

4. You use `CFrame.new()` to create a blank Cframe.

Exercises

Randomly spawn a model like a weapon or a health pack in a new location every 5 seconds, as shown in Figure 14.15. Here are some tips:

▶ Create individual variables for X, Y, and Z.

▶ You can use `random()` to return a random number. For example, `random( 1, 99)` produces a random number from 1 to 99.

▶ You can move the same model from place to place, or you can clone a new model from ServerStorage.

FIGURE 14.15
RBR1 health pack.

For this next exercise, make a platform that doesn't move back and forth on its own but only moves into place when a player clicks a switch, as shown in Figure 14.16. Use these tips:

▶ TweenInfo needs to be changed so that it does not repeat automatically.

▶ You can call `tween:Play()` from within a function.

FIGURE 14.16
A moving platform with a switch.

For the last exercise, create a block that spawns coins when players hit a block. Use these tips:

▶ You can place the coin object in replicatedStorage and spawn it using instance.new and parented to the block.

▶ Refer to Hour 4 for information about code that makes things happen when a part is touched.

▶ Try making the coin spin and bounce as it spawns using a tween.

HOUR 15
Sounds and Music

What You'll Learn in This Hour:

▶ How to create a soundtrack
▶ How to import music and sound assets
▶ How to create ambient sounds
▶ How to use code to trigger sounds
▶ How to group sounds

Music and sounds are useful tools for giving information to your players. Not only do audio elements set a tone for your game, but they can also let players know what's happening. For example, you can use sounds to let players know a GUI button is working or that an attack was successful.

This hour covers how to create a soundtrack for your game and how to use and import sound assets to give players audio feedback as they interact with the GUI and game elements.

Creating a Soundtrack

Music controls the mood of your game. You can use it to soothe players, get them excited, or reinforce the thematic elements of your game. If you're not a music composer, Roblox has a library of free music tracks that you can import into your game. Take a moment to browse Roblox's music library and then use the following steps to import music files into your game:

1. In the Toolbox, click the Marketplace tab and select Audio (Figure 15.1).

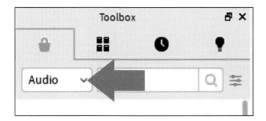

FIGURE 15.1
The Audio option on the Marketplace tab.

2. Click the search options icon (Figure 15.2) to find longer sound assets that would be appropriate for a soundtrack. You can also search by Roblox as a creator to find quality assets.

FIGURE 15.2
Searching for soundtrack options.

3. Click the Play button in the bottom-right of the icon (Figure 15.3) to preview the sound. If you like it, double-click to add it to your game.

FIGURE 15.3
The Play button.

4. In Properties, click Playing and Looped (Figure 15.4) to have the music continuously play.

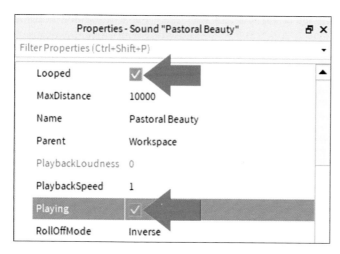

FIGURE 15.4
Create a loop to play music continuously.

Importing Music and Sound Assets

If you don't find music or sounds in the catalog that you want to use, you can upload your own music files. You pay a small Robux fee that covers the time it takes moderators to review every sound file that users upload. The cost is based on the total length of the sound asset. Audio files must be .mp3 or .ogg format, shorter than 7 minutes, and smaller than 19.5 MB. Table 15.1 shows the cost levels.

TIP

Make Sure You Have Rights

Keep in mind that uploading and/or using audio files that you don't have the rights to use is against Roblox's terms of service.

TABLE 15.1   Audio Upload Costs

Length of Audio File	Cost in Robux
0–10 seconds	20
10–30 seconds	35
30 seconds–2 minutes	70
2–7 minutes	350

To upload audio on the Roblox site, do the following:

1. Go to https://www.roblox.com/develop and click Audio.

2. Click Choose File and browse to the location of the file you'd like to upload. Click Estimate Price (Figure 15.5) to see the cost and then click the Purchase button to upload the audio file.

Create an Audio Don't know how? Click here

Audio uploads must be less than 7 minutes and smaller than 19.5 MB.

Find your .mp3 or .ogg file: [Choose File] No file chosen

Audio Name: [   ]

[Estimate Price]

FIGURE 15.5
Uploading an audio file.

3. Once the file is uploaded, it appears in the list on the page. After it's gone through moderation, it'll also be in the Toolbox under My Audio (Figure 15.6).

Audio ☐ Show Archived

MySound
Created 10/13/2020

FIGURE 15.6
Uploaded audio file in the Toolbox.

4. Click the name of the audio file to open its dedicated page and copy/record its numeric ID. You'll need this ID for playback testing in the sections later in this hour. In the following example, the numeric ID is 1837103530:

```
https://www.roblox.com/library/1837103530/Lucid-Dream
```

Creating Ambient Sounds

In addition to full-length background music, you may want some sounds that players hear only while they are in a certain area of the game. Think of anvil sounds coming from a blacksmith shop or NPCs murmuring to themselves.

The game developer typically needs to upload ambient sounds. To make your own ambient sounds, you can record real-life sounds using your phone or other recording equipment. Alternatively, you can search for free sound-generating programs online.

When you've created or found a file for an ambient sound you want to use, follow these steps:

1. Import the sound.

2. Place a part where you want the sound to play. This example uses a waterfall (Figure 15.7), so we're placing the sound at the bottom of the fall.

FIGURE 15.7
A waterfall that will include a sound file.

3. Insert a Sound object into the part (Figure 15.8).

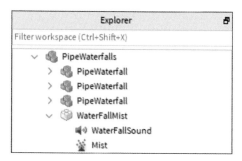

FIGURE 15.8
Inserting a Sound object into the appropriate part.

4. In Properties, scroll down to SoundId and enter the desired numerical ID, as shown earlier in the hour (Figure 15.9).

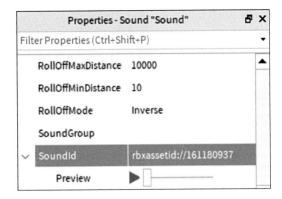

Properties - Sound "Sound"	
Filter Properties (Ctrl+Shift+P)	
RollOffMaxDistance	10000
RollOffMinDistance	10
RollOffMode	Inverse
SoundGroup	
∨ SoundId	rbxassetid://161180937
Preview	▶

FIGURE 15.9
SoundId property where the desired numerical asset number can be inserted.

5. Remember to enable Playing and Looped. Play with the following settings to control the range over which the sound can be heard:

   ▶ **MaxDistance:** Set the distance in studs over which the sound can be heard.

   ▶ **RollOffMode:** Set to InverseTapered to make it so the sound softens as you move away.

   ▶ **Volume:** Adjust so the volume is at the proper level. Be careful not to make every sound blaringly loud.

Triggering Sounds Using Code

You also can use code to control sounds if you need to set when the sound plays or change it depending upon what a player does. A common use of needing to call a sound is when players click something. The following steps explain how to add sound that plays when the player clicks a switch:

1. Use your switch setups from Hour 14 or create a part and insert a click detector. Use the following code to detect the click:

```
local button = script.Parent
local clickDetector = button.ClickDetector


local function onClick()
    print("button was clicked")
```

```
        button.CFrame = button.CFrame * CFrame.new(0, -.4, 0)
    end

clickDetector.MouseClick:Connect(onClick)
```

2. Insert the sound you want to use into the part. Here, we're using Button by Roblox (rbxassetid://12221967).

3. Use the highlighted code to get and play the sound:

```
local SoundService = game:GetService("SoundService")

local button = script.Parent
local clickDetector = button.ClickDetector
local buttonSound = button["button.wav"]

local function onClick()
    print("button was clicked")
    button.CFrame = button.CFrame * CFrame.new(0, -.4, 0)
    buttonSound:Play()
end

clickDetector.MouseClick:Connect(onClick)
```

Grouping Sounds

To make creating soundscapes easier, you can group sounds so the volume of the group can be turned up or down all at once. For example, one group of sounds might be what you want the player to hear in a nighttime campfire scene (Figure 15.10), and a different group to be heard in the morning.

FIGURE 15.10
Add nighttime sounds to a campfire scene.

To create and assign sound groups, use the following steps:

1. In SoundService, add a SoundGroup and rename it (Figure 15.11).

FIGURE 15.11
A renamed SoundGroup.

2. Select the sound you want to add to the group, and in Properties (Figure 15.12), click the field next to SoundGroup. Then click the desired group.

3. Repeat these steps to add all the sounds you want to include in the group.

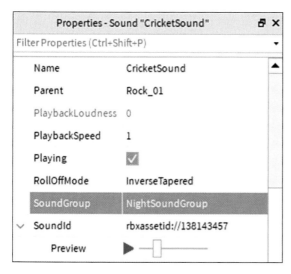

FIGURE 15.12
Assigning a sound to a SoundGroup.

Summary

Sounds can be used separately or as groups to make a game world feel more immersive and to provide players confirmation when they've interacted with objects. If you aren't a composer or audio designer, there's lot of sound and music files available from both Roblox and the community that you can use.

Each sound in the catalog has a unique numerical ID. In a Sound object, set the SoundId property to the numerical ID of the asset you would like to use. You can upload your own music and sounds for a small fee that covers the cost of moderation.

Q&A

Q. Do sounds need to be uploaded to a group for group games?

A. No. Unlike animations, any uploaded sound can be used in any game.

Q. Is it okay to use popular music within my game?

A. Like any type of asset, you should only use music and sounds that you have the rights to use. Do not use an artist's music without permission.

Q. Where can I find free music?

A. You can look for music uploaded by Roblox, or you can take advantage of free music creation tools to make your own.

Q. **How exactly do sounds make games more accessible?**

A. Your players will not all have perfect vision. Adding sound cues makes it so that all players will be less reliant on seeing every detail of your game to know what's happening.

Workshop

Now that you have finished, review what you've learned. Take a moment to answer the following questions.

Quiz

1. True or False: Different sounds should be used for different attacks.

2. True or False: You need a credit card in order to upload sounds.

3. If the sound should get quieter as the player moves away, change the _____ property to _____.

4. The farthest distance away from the source a player can possibly hear the sound is set by the _____ property.

5. True or False: All sounds should be at the exact same volume.

Answers

1. True. Ideally you want each player action to have a distinct sound.

2. False. The cost of moderation for sounds is paid for in Robux.

3. To make a sound quiet as the player moves away, change the RollOffMode property to InverseTapered.

4. You use the MaxDistance property to set the farthest distance away from a source where the player can still hear the sound.

5. False. Be careful to balance the sounds of your game so that it doesn't become overwhelming for players.

Exercises

Not only does sound make a world feel more alive, but it also improves the playability of your game. Sounds can be used to let players know they've clicked a shop button, run out of ammo, or that a threat is nearby. When possible, all of the different actions in your game should have their own unique sounds.

For the first exercise, go through your game and identify the different actions a player can take that don't already have sounds. This might be different attacks, earning points, or discovering treasure. Add sound to at least three different actions.

A soundscape is all of the little sounds around you that give you subtle clues about your environment. For the second exercise, think of three different places you visit and identify three unique sounds that you would hear in each area. For example, on a city block, you might hear car horns honking, food vendors yelling, crosswalk signals beeping, pigeons fighting, and airplanes flying.

When you're done, think about where you can create different soundscapes in your environment. Remember to adjust the Volume and MaxDistance so the sounds don't become too overwhelming.

One way to set the time of day in a Roblox game is by using `Lighting.TimeOfDay`. For example, 1:00 a.m. would be `Lighting.TimeOfDay = 01:00:00`.

The 24-hour clock must be set using the hours:minutes:seconds format `00:00:00`. 1:00 p.m. would be `13:00:00`.

For the last exercise, see if you can code a day/night cycle using a loop. Then, use the volume property on SoundGroups to create a different soundscape for the day (Figure 15.13) and night environments.

FIGURE 15.13
A daytime camp scene.

HOUR 16
Using the Animation Editor

What You'll Learn in This Hour:

▶ What the Animation Editor is
▶ How to create poses
▶ How to use Animation Editor tools
▶ How to work with animation events
▶ How to save and export animations

This hour introduces you to the Animation Editor, a must-have for adding detail to your game. Animation is perhaps one of the strongest tools for communicating action to your players. Without animation, the player can't clearly grasp what's happening; without good animation, the player is left confused or just underwhelmed.

Whereas bad animation can hinder the player's experience, great animation can make a game more immersive and memorable (Figure 16.1). It can clearly communicate action in a way that's less invasive than a UI pop-up. With the Animation Editor, you have a powerful built-in tool to create and upload custom animations to make your games unforgettable.

FIGURE 16.1
Animations abound in *Build It, Play It: Island of Move* by Roblox Resources.

Introduction to the Animation Editor

The Animation Editor is a tool that allows developers to create animations for characters, non-player characters (NPCs), and anything else they may want to animate in their games. Like most 3D animation software, the Animation Editor uses keyframes—timeline markers that indicate the beginning and ending of a pose—to construct movement. The developer creates the keyframes so they consist of core poses. Then the editor smoothly transitions from pose to pose, making for seamless animation.

Understanding Model Requirements

The Animation Editor can support a variety of models, as long as they are connected together as a rig with Motor6D's and contain a PrimaryPart. If you aren't familiar with rigging custom models together, you can insert a default R15 model because that is what we've used for this chapter's exercise.

To begin, open a separate baseplate, just to be extra safe and have your animations in a separate spot. Then follow these steps:

1. In the Plugins tab, click Build Rig (Figure 16.2).

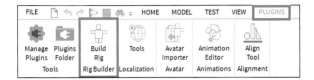

FIGURE 16.2
The Rig Builder on the Plugins tab.

2. Select Block Rig for this example (Figure 16.3).

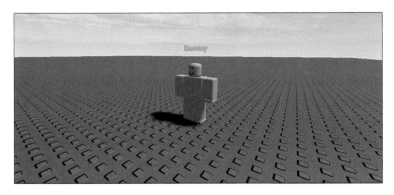

FIGURE 16.3
Inserted Block Rig R15 dummy.

With the test rig in your workspace, you're ready to open the Animation Editor.

Opening the Animation Editor

1. Open the Animation Editor, which is also located under the Plugins tab (Figure 16.4).

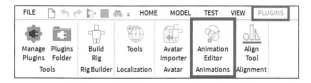

FIGURE 16.4
Path to find the Animation Editor.

2. Click your rig in the Animation Editor window (Figure 16.5) to select it. If prompted, name your animation and click Create.

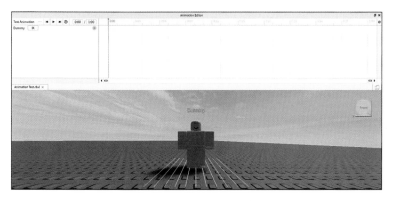

FIGURE 16.5
The Animation Editor.

Creating Poses

To animate your rig, you need to first define poses by moving specific parts, like the head or right leg, into the positions appropriate to the gesture you want to create. For example, if you want your rig to kick its right leg up, you need two poses: the leg starting on the ground, and the leg raised 90 degrees (Figure 16.6).

FIGURE 16.6
To tell the Animation Editor where each pose should begin and end, you create keyframes.

As mentioned earlier, a keyframe is a timeline marker that designates the beginning and end of a pose. To create a new pose using keyframes, use the following steps:

1. Move the scrubber bar (the blue line in Figure 16.7) to the time where you want to set the pose.

FIGURE 16.7
The scrubber bar in the Animation Editor.

NOTE

Timeline Units

By default, timeline units are expressed as seconds:frames and animations run at 30 frames per second (FPS), so 0:15 indicates ½ second. You can change this default setting by adjusting the Frame Rate in the Settings bar.

2. Click the part on your rig that you want to move.

3. Move and rotate the part to your desired orientation. When you do so, a track is created, and a new keyframe is created along the timeline, indicated by a diamond (Figure 16.8).

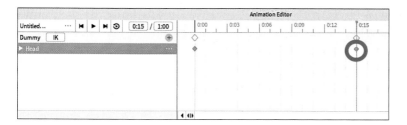

FIGURE 16.8
Creating a new keyframe.

NOTE

Toggling Move and Rotate Tools

When you're setting poses, you can toggle between the Move and Rotate tools by pressing Ctrl+2 or Ctrl+4 (Command+2 or Command+4 on a Mac), respectively. These modes work exactly like moving or rotating objects.

4. Continue moving or rotating parts until you get the desired pose. Whenever you adjust a specific part, a keyframe is defined for that part at the selected time.

5. When you're ready to preview the animation, click the Play button in the Animation Editor (Figure 16.9). You can also play or pause animations by pressing the Spacebar.

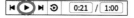

FIGURE 16.9
Click the Play button to watch your animation.

Once your basic poses are set, you may want to fine-tune individual keyframes to give your animation more polish. Here is a quick list of common keyframe manipulations:

▶ **Adding Keyframes:** Move the scrubber bar to a new position, click the ellipsis (…) button for a track, and select Add Keyframe.

▶ **Deleting Keyframes:** Select a keyframe and press the Delete key, or right-click and select Delete Keyframe from the context menu.

▶ **Cloning Keyframes:** Press Ctrl + C (Cmd + C for Mac) while highlighting your desired key-frames, and then press Ctrl + V (Cmd + V for Mac) after moving your scrubber bar to your desired pasting position.

▶ **Moving Keyframes:** Drag your keyframes to the desired spot on the timeline.

▼ TRY IT YOURSELF

Create an Attack Animation

How a character moves can give a lot of information about who they are. Big monsters might be slow and drag their feet as they take a swipe. Young anime protagonists might jump and spin as they punch. Try creating a distinctive attack animation.

1. In the middle of the timeline, create the pose for the finished attack (like the one in Figure 16.10). You should see keyframes appear at both the cursor and the beginning of the time-line, as shown in Figure 16.11.

FIGURE 16.10
A pose for the finished attack.

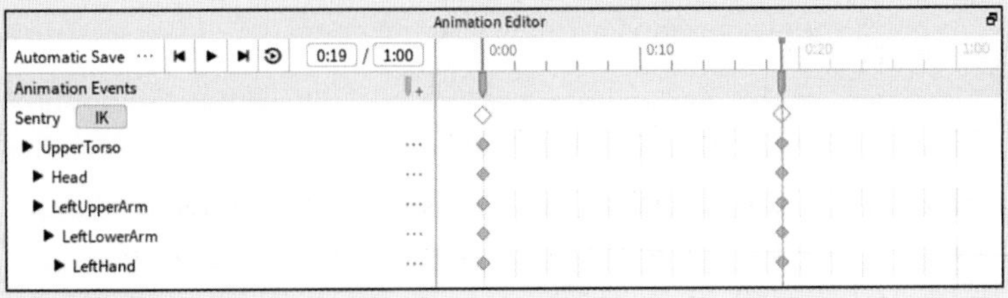

FIGURE 16.11
Keyframes for the pose.

2. Click Play to see how your animation looks.

3. Drag the keyframes to make the animation faster or slower, and add new keyframes where needed to make the animation smoother (Figures 16.12 and 16.13).

FIGURE 16.12
From left to right, positions of starting, winding back, and swiping.

FIGURE 16.13
Keyframes displaying the various positions.

Saving and Exporting Animations

To save an animation, use the following steps:

1. Click the ellipsis (…) at the top right of the Animation Editor.

2. Select Save As or Save depending on whether you want to create a new animation object or update a preexisting one. This will save your new animation in a model named AnimSaves (Figure 16.14) in the dummy you animated.

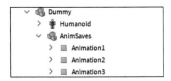

FIGURE 16.14
Where to find your saved KeyframeSequences.

NOTE

Saving Versus Exporting

The preceding steps won't save your animation to Roblox's servers; they only keep it in the model. To save an animation in Roblox and use it, you have to export it, as described in the next steps.

To export animations for proper in-game use, use the following steps:

1. Click the ellipsis (…) at the top right of the Animation Editor.

2. Click Export.

3. Give your animation file a name, description, and a group owner.

4. Click Submit.

To copy your animation ID for use, click the Copy button right next to the animation ID. To find your animation ID later, do the following:

1. Click the ellipsis (…) at the top right of the Animation Editor. Then select Import, From Roblox.

2. Select the animation and then copy the ID number at the bottom (Figure 16.15).

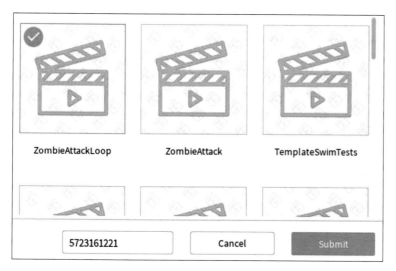

FIGURE 16.15
Copying the animation ID.

Easing

Hour 15 introduced easing direction and styles. These tools determine how your keyframes move from one pose to another. They are essential because they can easily help you create intricate, life-like animations in less time.

By default, a part moves from one keyframe to the next in the steady motion of linear easing. However, you may want to customize your easing to make the animation more dynamic. To change easing for one or more keyframes, select the keyframes you want to modify, right-click, and select your options from the Easing Styles and Easing Directions menus. You can then edit the style and directions to change how the keyframe interpolates from one pose to another.

TRY IT YOURSELF ▼

Easing Styles

Experiment with different easing styles on your attack animation.

Working with Inverse Kinematics

Inverse Kinematics (IK) is a widely useful tool for quickly being able to position joints. A good example of IK is keeping the legs in place to calculate the position of multiple joints by just moving a single joint. For instance, to create a crouch (Figure 16.16), you could move just the lower torso downward and keep the feet in place.

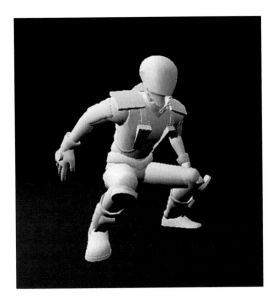

FIGURE 16.16
An example of Body Part IK on the LowerTorso.

If you animate only the LowerTorso, the calculations for the legs are automatically handled.

Enabling IK

To start animating using IK, click the IK button in the Animation Editor (Figure 16.17) to bring up the Manage IK window on the left of the screen. At the bottom of the window, click Enable IK.

FIGURE 16.17
The IK button.

Roblox IK can be categorized into two separate modes: Full Body and Body Part (Figure 16.18).

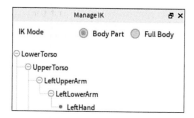

FIGURE 16.18
The two IK settings: Body Part and Full Body.

In Body Part IK mode, when you move a body part, only that part experiences any movement. For example, moving the right arm only affects parts that make up the right arm (Figure 16.19).

FIGURE 16.19
Body Part IK enabled.

In Full Body IK mode, when you move a body part, all parts of the rig will be considered. For example, moving the right arm calculates effects in the rest of the body (Figure 16.20). If you don't want certain parts to be calculated, you can pin them to keep them stationary (see the next section).

FIGURE 16.20
Full Part IK enabled.

Pinning Parts

With Full Body IK, you may want to pin parts so they don't move. Simply click the pin icon next to the name of the part you want to pin (Figure 16.21). In Figure 16.22, the feet of the rig are pinned.

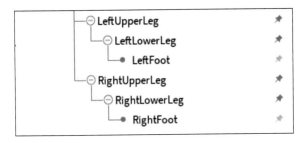

FIGURE 16.21
Click the pin icon next to a part.

FIGURE 16.22
R15 character rig with its two feet pinned.

Animation Settings

You will probably want animations such as attacks to play just once when triggered. In other cases, animations like a run cycle need to play over and over until something stops it. Some animations are also more important than others, and you will want to prioritize their playing over other animations that might be also be triggered at the same time.

Looping

You can enable looping for animations by toggling the Looping button (Figure 16.23). When this button is enabled, it will export the animation as a loop. Be careful, though: When it loops, it stops at the final keyframe's row in your timeline and then goes back to the beginning without any transition. To get around this, if you're making animations that need to loop, you can copy your first set of keyframes to the end of your timeline to get a seamless loop.

FIGURE 16.23
Loop enabled.

Priority

In a gameplay setting, different player states may require different animations. For example, a player's attack animation would be different than an idle animation. In most scenarios, you'll want the attack animation to have a higher priority than the idle animation so that the two actions won't conflict. Animation priority can be represented visually as shown in Figure 16.24.

FIGURE 16.24
Animation priority.

To view and adjust animation priority, click the ellipsis near the top right of the Animation Editor and select Set Animation Priority (Figure 16.25).

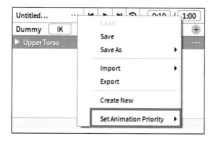

FIGURE 16.25
Setting animation priority.

In the window that opens, you can see your current animation priority and change it as needed. An important thing to remember is that the lower the priority, the more likely the animation is to be overwritten by an animation with a higher priority.

Working with Animation Events

When you're scripting animation, it's quite often that you'll want to give off a signal when a certain keyframe is reached. A common example is playing footstep sounds during a walk cycle.

To achieve this, you can add event markers, which are checkpoints that send off a signal when a certain animation spot is reached. You can then use `GetKeyframeMarkerReached()` to hear when that keyframe is reached.

To show the animation event track, you first need to enable it. Click the Settings icon and select Show Animation Events (Figure 16.26).

FIGURE 16.26
Animation Editor's settings.

After enabling Show Animation Events, your timeline should have a new track called Animation Events listed at the top of your previous tracks (Figure 16.27).

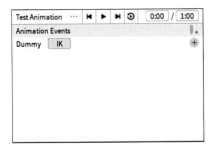

FIGURE 16.27
The new Animation Events track.

Adding Events

Adding animation events is a pretty simple process. Add an animation event to mark both the beginning and end of your animation. Later in this hour, you use these events in a script. Follow these steps:

1. Right-click the point in the timeline where you want the event, and select Add Animation Event Here.

2. Create two new events at the beginning and the end of the animation named AnimationStart and AnimationEnd.

3. Once you define the necessary fields, click Save. You see a new keyframe marker where your scrubber bar was last placed (Figure 16.28).

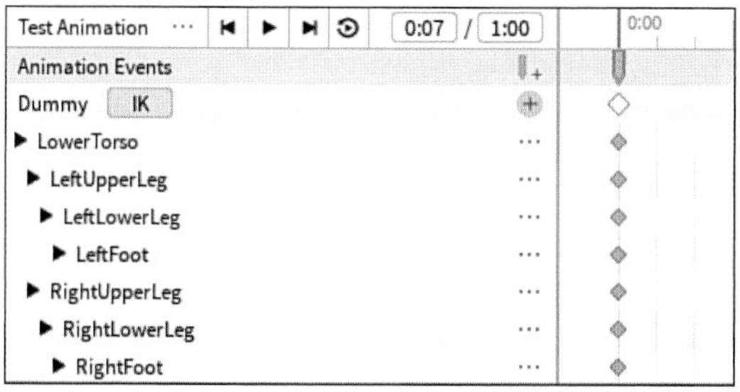

FIGURE 16.28
An animation event paired with keyframes.

For scripting, you can have additional parameters to pass to your `GetKeyframeReachedSignal()` event.

Moving and Deleting Events

If you ever want to move an event, it's as simple as dragging the animation event to a new position on the animation track. To delete events, you can press the Delete key on your keyboard or right-click an animation event and select Delete Selected from the context menu.

Cloning Events

As you create events, they become available for use throughout the animation, and you can clone them for reuse. For instance, you can create a HandWave event marker at the point where a character's right hand is raised, and then use the same event for waving the other hand.

To clone animation events, you first select the animation event you want to clone and press Ctrl + C (or Cmd + C on a Mac). Then move your scrubber bar to where you want to paste the animation event and press Ctrl + V (or Cmd + V on a Mac).

Implementing Events in Scripts

To implement animation events in a local script, you use a mixture of `GetMarkerReachedSignal()` and the animation object. For the following example, when a player presses F on the keyboard, the character stops and attacks. You need to create animation events for both the beginning and end of the animation.

1. Make sure your animation with events for AnimationStart and AnimationEnd has been saved and exported to Roblox.

2. Select StarterPlayer, StarterPlayerScripts and add a LocalScript named ZombieAttack.

3. Type the following code:

```
local UserInputService = game:GetService("UserInputService") -- Detecting
    player input
local Players = game:GetService("Players")                  -- Get local
    player
local player= Players.LocalPlayer                           -- Our player

-- Get the player's character model that needs to be animated
local characterModel = player.Character or player.CharacterAdded:Wait()
local humanoid = characterModel:WaitForChild("Humanoid")

local animation = Instance.new("Animation")        -- New animation object
animation.AnimationId = "rbxassetid://000000000"  -- Use your animation asset
    ID

local animationTrack = humanoid:LoadAnimation(animation)

-- Detects if player is pressing F and plays animation
UserInputService.InputBegan:Connect(function(input, isTyping)
    local normalWalkSpeed = humanoid.WalkSpeed
    if isTyping then return end

    if input.KeyCode == Enum.KeyCode.F then
        animationTrack:Play()
    end

    animationTrack:GetMarkerReachedSignal("AnimationStart"):
    Connect(function()
        humanoid.WalkSpeed = 0                     -- Stops the player from
        moving
    end)

    animationTrack:GetMarkerReachedSignal("AnimationEnd"):Connect(function()
        humanoid.WalkSpeed = normalWalkSpeed  -- Allows movement again
    end)
end)
```

This code detects when the player presses F, plays the animation, and stops the player from moving until the animation ends. If you've followed the steps properly, now when you press F, your very own custom attack animation will play!

Replacing Default Animations

By default, Roblox player characters include common animations like running, climbing, swimming, and jumping. However, at some point, you might feel like you want to replace these defaults with your own animations. Luckily, Roblox enables you to do this pretty easily through the use of scripts. If you have a valid animation ID, you can replace the default animations with your own custom ones.

The process of replacing default animations goes as follows:

1. Insert a script in ServerScriptService.

2. Type the following script but replace the animation IDs with an ID you've created. Also edit animations as needed.

```
local Players = game:GetService("Players")
local function onCharacterAdded(character)
    local humanoid = character:WaitForChild("Humanoid")

    for _, playingTracks in pairs(humanoid:GetPlayingAnimationTracks()) do
        playingTracks:Stop(0)
    end

    local animateScript = character:WaitForChild("Animate")
    animateScript.run.RunAnim.AnimationId = "rbxassetid://616163682"
        -- Run
    animateScript.walk.WalkAnim.AnimationId = "rbxassetid://616168032"
        -- Walk
    animateScript.jump.JumpAnim.AnimationId = "rbxassetid://616161997"
        -- Jump
    animateScript.idle.Animation1.AnimationId = "rbxassetid://616158929"
        -- Idle (Variation 1)
    animateScript.idle.Animation2.AnimationId = "rbxassetid://616160636"
        -- Idle (Variation 2)
    animateScript.fall.FallAnim.AnimationId = "rbxassetid://616157476"
        -- Fall
    animateScript.swim.Swim.AnimationId = "rbxassetid://616165109"
        -- Swim (Active)
    animateScript.swimidle.SwimIdle.AnimationId = "rbxassetid://616166655"
        -- Swim (Idle)
    animateScript.climb.ClimbAnim.AnimationId = "rbxassetid://616156119"
        -- Climb
end

local function onPlayerAdded(player)
    player.CharacterAppearanceLoaded:Connect(onCharacterAdded)
end

Players.PlayerAdded:Connect(onPlayerAdded)
```

Note that sometimes it takes an animation a bit of time to go through moderation. If you only just published the animation and it isn't working, try again in a minute.

Summary

In this hour, you've learned how animations can bring your characters to life and give them personality. You've learned how to use Inverse Kinematics (IK) and easing styles, and you've been introduced to animation priority, which is very important in making sure your animations play at the right time, and how it affects animations in a game.

You've also learned how to add animation events to your animations and to use those events to add custom functionality linked to your animations. This is a very useful tool for syncing effects with animations. Finally, you've learned how to save, export, and use animations in a proper script.

Q&A

Q. **Will animation events do anything if I don't connect them?**

A. No, animation events need to be connected in a script to have any functionality.

Q. **What if I want to edit animations I uploaded?**

A. To edit already uploaded animation, you can press the ellipsis at the top of your screen, press Import -> From Roblox, and then select the animation you want to edit. To update that animation, all you need to do is re-export it to the animation you were editing. It will automatically update all scripts that make usage of that animation.

Q. **My development partner made an animation, but I can't use it. Why?**

A. When it comes to animations, only the user who originally uploaded it can use it. To use it, you'll need a copy of the KeyframeSequence to upload yourself or they can republish the animation to a group you are both in.

Workshop

Now that you have finished, review what you've learned. Take a moment to answer the following questions.

Quiz

1. What plugin allows you to add R15 and R6 rigs to your game?

2. Saved rotation and position information within the Animation Editor timeline is called a _____.

3. True or False: To create an animation event, insert an AnimEvent object into the rig.

4. True or False: Toggling Looping on only loops the animation for testing purposes, not in game.

Answers

1. The Build Rig plugin allows you to add R15 and R6 rigs to your game.

2. Saved rotation and position information within the Animation Editor timeline is called a keyframe.

3. False. Animation events are created by right-clicking a keyframe and selecting "Add animation event here."

4. False. Toggling Looping on causes the exported animation to loop in game until it is interrupted.

Exercises

The exercises use the concepts you learned up to this point. If you get stuck, don't forget to refer to the previous pages!

In the first exercise, you replace the default animation with a custom animation and emit a particle effect.

1. Create a custom run animation and add animation events whenever the character's foot touches the ground.

2. Upload the animation, and grab the ID for later use.

3. Replace the default animation with your own, either by using the method from this hour or looking at the animate script that gets pasted in your character when you play the game.

4. Create a particle emitter, add a `GetMarkerReachedSignal()` connection, and then call `ParticleEmitter:Emit()` when the footstep animation event is reached.

5. Test it!

In this exercise, add a custom death animation that plays when the player's health reaches 0. For extra detail, you can add a custom death sound.

1. Create the custom death animation. Using animation events is optional.

2. Upload the animation and grab the ID for later use.

3. Create a new client-sided script, get the character, and get the character's humanoid.

4. Create a new animation object, add your animation ID to your new animation, and make a new animation track.

5. On the Humanoid.Died event, make a new function that plays your new death animation; also connect any animation events if you added them.

6. Test it!

Combat, Teleporting, Data Stores

What You'll Learn in This Hour:

▶ How to use tools
▶ How to build a basic fighting game
▶ How to use teleportation
▶ How to use persistent data stores

Some of today's most popular Roblox games are combat based. From sword-fighting tournaments to superhero worlds, these games offer players action-packed game mechanics. In this hour, you find out how to use tools to build a bare-bones fighting game. You also learn how to build swords using the Roblox tool container, how to teleport within and between servers, and how to save your score data between sessions.

Introduction to Tools

The main mechanic of any combat game is a player's tool, so let's understand this basic element before we build our game. Tools are objects that are built into the Roblox backpack system (Figure 17.1). They can be anything—from spells to swords—and they can easily be equipped in your game with little scripting knowledge.

FIGURE 17.1
Backpack with tools for building in *Roblox Battle Royale* by Roblox Resources.

This section explains how tools work and how you use code to make it so that the tool harms whatever player it touches. In later hours, you find out how to add animations for slashing and attacks.

Tool Basics

Tool objects act as containers to hold the various parts that comprise a tool, including meshes, scripts, sound effects, and value objects. When tools are placed within the player's backpack, icons appear in a hotbar at the bottom of your screen, as shown in Figure 17.2. You can equip them using assigned keybindings, which are numbered from 0 to 9.

FIGURE 17.2
Weapons backpack in *Roblox Battle Royale* by Roblox Resources.

Creating a Tool

You can place tools within StarterPack if you want them with the player's backpack at the start of the game. Alternatively, you can place them in the game world for the players to find and pick up.

In Explorer, insert a few tool objects into StarterPack and rename them so you can tell them apart, as shown in Figure 17.3.

FIGURE 17.3
Roblox toolbar with tool objects.

The tool object starts out as an empty container that can hold the images, models, and scripts that make a finished tool function. Playtest the game now, and you'll notice the tools have loaded into the player's backpack and are displayed as a UI at the bottom of the screen (Figure 17.4).

FIGURE 17.4
The tools in the player's backpack.

Later in the hour, you'll be able to activate items using the tool objects, either via the UI or via keybindings (0 through 9).

Tool Handle

To be held by the player, the tool object requires a part named Handle to mark where the player's hand grips the tool. The Handle part must be parented directly to the tool object. Without a part named Handle, the tool just spawns in its original position in the Workspace, which is default position [0,0,0] and does not move with the player.

NOTE

Tools Without a Mesh

If the tool does not need to be held, such as with the building system shown in Figure 17.1, uncheck RequiresHandle.

Use the following steps to create a handle:

1. Insert a part into the Workspace.

2. Shape and design the part into the handle you want.

3. Change the Name property to Handle.

4. Move the handle part into First Tool under StarterPack (Figure 17.5). Notice the handle disappears once it is no longer parented to Workspace.

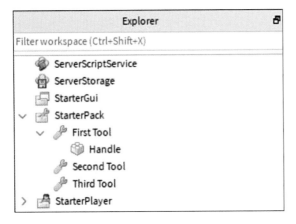

FIGURE 17.5
Handle part moved into First Tool within StarterPack.

5. Playtest, click the First Tool in the backpack, and the tool appears in your player's hand (Figure 17.6).

FIGURE 17.6
Tool handle.

Notice that the tool welds to your player's right hand. When a tool is equipped, the handle is welded to the player's right hand by default, but you can manually code a handle to weld to the left hand.

Tool Appearance

With the tool successfully created, you can use the tool's Appearance properties (Figure 17.7) to modify the tool's grip—that is, the direction the tool faces as it is being held by the player.

CAUTION

Don't Use the Rotation Tool

It's really important *not* to use the rotation tool for this as it won't work and can break the tool.

▶ **GripForward:** One of three properties denoting Orientation, specifically the ZVector (R02, R12, R22)

▶ **GripRight:** Orientation for the XVector (R00, R10, R20)

▶ **GripUp:** Orientation for the YVector (R01, R11, R21)

▶ **GripPos:** Handles positional offset from hand

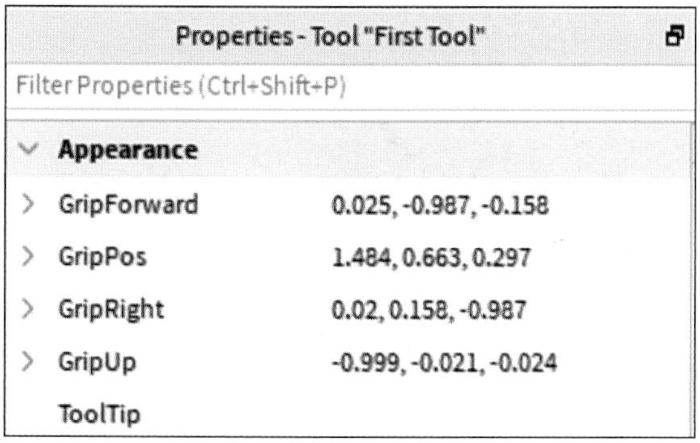

Properties - Tool "First Tool"	
Filter Properties (Ctrl+Shift+P)	
∨ **Appearance**	
〉 GripForward	0.025, -0.987, -0.158
〉 GripPos	1.484, 0.663, 0.297
〉 GripRight	0.02, 0.158, -0.987
〉 GripUp	-0.999, -0.021, -0.024
ToolTip	

FIGURE 17.7
Tool grip properties that control the direction a tool faces while being held.

Use a Plug-In

By far the easiest way to find the correct orientation settings is to use a helpful plug-in costing only 5 Robux developed by CloneTrooper1019. The plug-in comes complete with a separate visualization panel that displays how the handle looks when held, as well as intuitive rotation and translation tools (Figure 17.8).

To download this plug-in, go to https://www.roblox.com/library/174577307/Tool-Grip-Editor.
Once installed, the plug-in can be selected in the plug-ins tab.

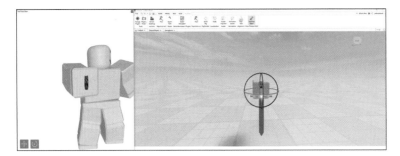

FIGURE 17.8
The Tool Grip Editor plug-in.

Equipping the Tool In-Game

Alternatively, to find out what settings to use for the Tool's grip properties, you may equip the tool
in-game and then change the values so you can see the direct results. Once the correct property values are found, they can be copied or written down. To equip it in-game, you can do the following:

1. Start the playtest.

2. Select the tool. Unequipped tools can be found Players, Player Name, Backpack (Figure
   17.9). Equipped tools can be found under Workspace, Player Name, Tool Name.

FIGURE 17.9
How to find unequipped tools while playtesting.

3. Experiment with the properties until the tool is positioned how you want. Do not use the
   rotation or scale tools.

4. Copy the tool (Ctrl+C/Cmd+C) before exiting the playtest session so that you don't lose
   your work.

5. Once you stop the playtest, you can paste the tool back into the StarterPack and delete the old version of the tool.

Making a Fighting Game Sword

Now that you understand how the tool object functions and how to modify it, you can implement what you've learned into your game. Open a place file and start working!

First, you need some form of sword model as your tool. Make the kind of sword you'd like using parts, unions, or meshes. Remember to parent the handle directly to the tool object. If you're using more than one part or object, make sure additional parts are all welded to the handle to hold them together.

Creating the Script

First, you need to set up your script object by doing the following:

1. Insert a script directly under the tool container.

2. Name the script so you remember what it does in the future—for instance, SwordController.

3. Copy the following script so that players are harmed when attacked with the sword:

```lua
local COOLDOWN_TIME = 0.5
local DAMAGE = 30

local Players = game:GetService("Players")
local ServerStorage = game:GetService("ServerStorage")

local tool = script.Parent
local swordBlade = tool.Handle -- Change to match your sword blade

local humanoid, animation, player

local canDamage = true
local isAttacking = false

local function onEquipped()
-- Set humanoid, animation, and player variables
local character = tool.Parent
humanoid = character:WaitForChild("Humanoid")
animation = humanoid:LoadAnimation(tool:WaitForChild("Animation"))
player = Players:GetPlayerFromCharacter(character)
end

local function onDetectHit(otherPart)
local partParent = otherPart.Parent
local otherHumanoid = partParent:FindFirstChildWhichIsA("Humanoid")
-- Make sure blade doesn't harm controlling player
```

```
if otherHumanoid and otherHumanoid == humanoid then
return
-- Checks that there's another player and attack isn't on cooldown
elseif otherHumanoid and isAttacking and canDamage then
canDamage = false
otherHumanoid:TakeDamage(DAMAGE)
end
end


local function onAttack()
local waitTime = math.max(animation.Length, COOLDOWN_TIME)
if not isAttacking then
isAttacking = true -- Disable repeated attack attempts during animation
canDamage = true
animation:Play() -- Play animation
wait(waitTime) -- Wait until animation is complete or cooldown is finished,
whichever is longer
isAttacking = false -- Re-enable attack attempts
end
end


tool.Equipped:Connect(onEquipped)
tool.Activated:Connect(onAttack)
swordBlade.Touched:Connect(onDetectHit)
```

Modifying This Script to Work for Your Game

There are a couple steps required to make the script work with your specific weapon and game. Once you customize the script, you can test it by using the Network Simulator to simulate multiple players.

NOTE: This script assumes your game has attack animations such as those created in Hour 16, "Using the Animation Editor," to use with the sword. If you decide not to use an animation, make sure to remove any lines with `animation`; otherwise, it will error.

To include an animation, do the following:

1. Find the ID of the desired animation, as shown in Hour 16.

2. Insert an Animation object into the tool (Figure 17.10).

FIGURE 17.10
Hierarchy showing inserted Animation object.

3. In Properties, AnimationId, paste the asset ID and press Enter.

Handle

Remember to modify the variable `local swordBlade` to reference the part of the weapon that does damage. If your sword is all one mesh, or part, it should look like local `swordBlade = tool.Handle`. If your model has multiple parts, as in Figure 17.11, the code snippet would look like the following:

```
local swordBlade = tool.Sword.Blade
```

FIGURE 17.11
Sword container.

Testing So Far

At this point, you should be able to attack other players but not yet earn points. Use the Network Simulator to simulate multiple players in the game and test the script. Go to the Test tab (Figure 17.12), and in the Clients and Servers section, select the desired number of players to simulate. Click Start to begin testing, and Cleanup when you are ready to stop.

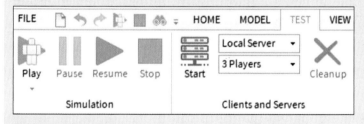

FIGURE 17.12
The Network Simulator.

Figure 17.13 shows a player holding a sword.

FIGURE 17.13
The finished product.

Additional Properties

Tools have built-in functionality that improve their behavior and aesthetics to help bring your game together:

1. **TextureId:** Instead of displaying a tool name in the UI bar at the bottom, set an image.

2. **ToolTip:** Allows you to display a tip when the player hovers over the tool icon in game.

3. **CanBeDropped:** When enabled, the player can remove the tool from their backpack by pressing Backspace.

4. **Requires Handle:** If your tool does not require parts, meshes, or unions, you can deselect this, which is useful for spells that don't require the tool to be physically connected to the player's character.

5. **Tool.ManualActivationOnly:** A really useful tool for stopping client clicks from firing Activated.

Teleportation

In our fighting game, we want to build a fighting area for combat and a lobby area where the players can rest between sessions. To move players between these areas, we use teleportation.

Roblox offers two types of teleportation: between servers and games and within the same server. Teleportation within a place uses CFrames, much like you would move any other object with a couple differences. Teleportation between different places or servers uses Roblox's aptly named TeleportService and its related APIs to move between servers and game universes.

Table 17.1 shows possible use cases for both teleporting players using CFrame within a place and using TeleportService to move players between places and servers.

TABLE 17.1   Game Design Use Cases

Within a Place	Between Places and Games
Event- or competition-based games.	Need to set up a new game environment.
	The game is too large—for example, ultimate Driving.
Move the player over large areas quickly.	Private servers.
Move the player between areas they otherwise may not be able to access, such as VIP rooms.	
Quick player transfer required.	

Teleporting Within a Place

Teleporting within a place allows you to move the player between areas of the map with ease. This is helpful for going through dud doors or teleporting the player between events. Figure 17.14 shows a player teleporting through a portal.

FIGURE 17.14
Portal.

One of the main concerns you run into when attempting to teleport a player's character is moving all the joints in one go without killing the player. The player dies if the torso/UpperTorso or head is displaced from the rest of the body. You can test this in play mode by using the move tool on a player joint.

To avoid this issue, use the CFrame property found on the HumanoidRootPart property found on all characters because HumanoidRootPart is not just the movement controller of the player characters; it's also the root part. Consequently, you can manipulate its position, directly affecting all limbs and connected objects.

▼ TRY IT YOURSELF

Teleporting to Fighting Game Areas

In this example, you create a quick-fire game where the player gets a certain amount of time to kill as many people as possible in the arena before getting transferred back to the lobby for the next game. The following code is very simple and can be expanded upon.

The lobby and arena will be within the same place, so use the CFrame method. Feel free to make this as complex or as simple as you want. We also need the CFrames of the spawning positions so the script knows where to place people.

To teleport players within a single place, insert a new script into ServerScriptService and add the following code:

```
local Workspace = game:GetService("Workspace")
local Players = game:GetService("Players")

local ARENA_CFRAME = CFrame.new(0, 100, 0)
local LOBBY_CFRAME = CFrame.new(100, 100, 100)

local TELEPORT_COOLDOWN = 0.5
local AREA_COOLDOWN =  30

local newArea

local function TeleportAllCharacters(location)
    for _,player in ipairs(Players:GetChildren())do
        local character = player.Character or player.CharacterAdded:wait()
        local humanoidRootPart = character.HumanoidRootPart
        humanoidRootPart.CFrame = location
        wait(TELEPORT_COOLDOWN)
    end
end

-- Teleports players back and forth
while true do
    TeleportAllCharacters(newArea)
    if newArea == ARENA_CFRAME then
        print("Players teleported to Arena")
        newArea = LOBBY_CFRAME
    else
```

```
        print("Players teleported to Lobby")
        newArea = ARENA_CFRAME
    end
    wait(AREA_COOLDOWN)
end
```

The preceding script allows for 30-second games where the players fight to the death before being teleported back to the lobby. Our tools are placed in StarterPack, so they'll automatically be replaced on the player when they respawn. If you want the player to lose the tool when they head back to the lobby, you could either remove it via script or create two teams: Lobby and Arena. Then, if you place tools within the arena team object, they'd be placed within the player's backpack.

TIP

Using a Part to Determine CFrame Coordinates

To easily find the correct CFrame coordinates, place a part into the world, position it in the desired location, and then copy the part's position value into ARENA_CFRAME or LOBBY_CFRAME. Make sure to include CFrame.new().

The script is split into three sections—Variables, TeleportAllCharacter, and Loop—each with a distinct purpose:

- ▶ **Variables:** Set up references for the rest of the script, including the locations that the players will be teleported to.
- ▶ **TeleportAllCharacter:** Goes through all player characters and sets the CFrame of the Humanoid Root Part (HRP).
- ▶ **Loop:** Runs constantly in the background, every 30 seconds changing the area before teleporting players.

Teleporting Between Places

Teleporting between places uses a vastly different system compared to within the same place. This is because you're having to load up a totally new environment for the player, which means contacting Roblox's services and asking them to teleport and load the player into new place files.

Game Universes

As mentioned in the beginning of the hour, Roblox games are made up of one or more places. This must include a Start Place, which acts as the landing bay for anyone who clicks your game. With Roblox's teleport service, players can transport between any places within the same game or to the Start Place of a different game. Figure 17.15 shows where players in game A can be teleported to.

FIGURE 17.15
Teleporting rules—a check mark indicates places available for teleportation.

TeleportService

TeleportService works between game universes and within them to transport players between servers and places. It offers a range of functions, including the ability to teleport individuals and groups of people, and is responsible for transporting players between place files within the same game. This is extremely useful for larger games or games with different levels. To reference TeleportService, we use the service provider function GetService() like so:

```
local TeleportService = game:GetService("TeleportService")
```

NOTE

TeleportService in Studio

TeleportService does not work in Studio, so publish your game and test within a live server.

Functions

The following are the most common functions you'll use with TeleportService:

- ▶ Teleport(placeId , player , teleportData , customLoadingScreen)
    - ▶ placeId: The ID of the arriving place file.
    - ▶ player: The player Instance—that is, game.Players.LocalPlayer.
    - ▶ teleportData: A list/array containing any parsed additional data such as previous placeId.

▶ customLoadingScreen: Passes a GUI (for example, ScreenGui) to the target location, which can then be used as a loading screen for the transition.

▶ GetArrivingTeleportGui(): If a GUI was sent alongside the Teleport() function, it will allow the new "place" to use it.

▶ TeleportPartyAsync(...): Teleport() but for multiple players.

▶ ReserveServer(placeId): Returns an access code to be used with TeleportToPrivateServer.

NOTE

List of Functions

A comprehensive list of functions can be found in the API documentation.

Many functions from TeleportService can be used from both the client and server. Client-sided functions do not require a player argument to be specified.

Grabbing the PlaceId

To teleport the players to a place, you need to get the ID. Open the Asset Manager that you previously used for uploading images and meshes. Once it's open, right-click the relevant place and select Copy ID to Clipboard (Figure 17.16).

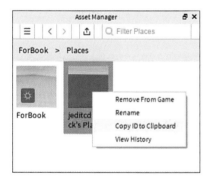

FIGURE 17.16
Copying the ID to the clipboard.

Client Example

Let's look at a working demonstration of a client-side teleport function to see how to teleport the client from a LocalScript in StarterPlayerScripts.

```
local TeleportService = game:GetService("TeleportService")
local PLACEID = 1234567 --Make sure to replace this with your PlaceID
local WAIT_TIME = 5 -- In Seconds, how long until player teleports to PLACEID
wait(5)
TeleportService:Teleport(PLACEID)
```

Although you can specify a `customLoadingScreen`, it's recommended that you instead use the `:SetTeleportGui(GUI)` function because it allows you to assign a UI element before the teleport function. You can then use `:GetArrivingTeleport Gui()` on the arriving server:

```
local TeleportService = game:GetService("TeleportService")
local PLACEID = 1234567 --Make sure to replace this with your PlaceID
local WAIT_TIME = 5 -- In Seconds, how long until player teleports to PLACEID
local GUI = game.ReplicatedStorage:WaitForChild("ScreenGui") -- New Line Of Code

wait(5)

TeleportService:SetTeleportGui(GUI) --New Line Of Code
TeleportService:Teleport(PLACEID)
```

To test this script, publish it and try it in-game via the website because teleporting doesn't work within Roblox Studio. (Place files can be opened by hand via Game Explorer.)

Server Example

Now let's take a look at teleporting from the server side. This method offers a couple advantages, such as teleporting groups of players. It also offers security because a client-side script could be removed by an exploiter, stopping the exploiter from being teleported. This script should be placed within ServerScriptService:

```
local TeleportService = game:GetService("TeleportService")
local Players = game:GetService("Players")

local placeId = 000000000 -- Replace with destination place ID
local SESSION_TIME = 30 -- How long before teleport

-- Teleports all players to a new place
local function teleportAllCharacters(location)
    for allPlayers, player in ipairs(Players:GetChildren())do
        TeleportService:Teleport(placeId, player)
```

```
        end
end

wait(SESSION_TIME)

teleportAllCharacters()
```

TIP

Wrapping Teleport Functions in pcall

Later in the chapter, we demonstrate why you should wrap your teleport functions in a pcall in case they fail, which can happen for a variety of reasons. That way, it fails safely and lets you know so you can call the function again. We discuss pcall in more detail later in this hour in the "Protecting and Responding to Errors" section.

While this covers the main uses of `TeleportService`, another way to use it is to set up private matches and so on. Using `TeleportService:ReserveServer()` coupled with `Teleport-ToPrivateServer()` allows you to open a reserved server. It works much like the purchasable Private servers, but only the developer on the server side can initiate them.

TRY IT YOURSELF ▼

Teleporting to Fighting Game Areas in Different Places

While short burst games require the quick load time that is only possible within a server, you can use teleportation for larger arena sessions or where the environments and functions of each section have little crossover. As with last time, the first step is to set up arena and lobby areas (Figure 17.17), but this time you do it on different place files.

FIGURE 17.17
Asset Manager.

Make sure the lobby is set as the Start Place; otherwise, players will load into the arena, which could possibly disturb a game in process.

Once complete, you need to place one or more spawn locations on each place file for the players. You also need to place your weapon in the arena StarterPack but not the lobby because you're in completely different servers with a fresh environment. The tools don't transfer, which removes the risk of players having the weapon in the lobby.

Because this game is made up of two place files with two separate environments (two separate SeverScriptService, Workspace, and so on), you need two scripts in each. Place each within ServerScriptService to limit abuse and so you can teleport all players at once between places (and servers), like so:

```
local TeleportService = game:GetService("TeleportService")
local Players = game:GetService("Players")

local PLACE_ID = 00000 -- Replace with place Id
local SESSION_TIME = 30 -- How long before teleport

-- Teleports all players to a new location within the place
local function teleportAllCharacters(location)
      for allPlayers, player in ipairs(Players:GetChildren())do
            TeleportService:Teleport(PLACE_ID, player)
      end
end

wait(SESSION_TIME)

teleportAllCharacters()
```

Once complete, make sure both places are published and to match the PLACEID to the corresponding placeID; for instance, if you're currently in the lobby, the placeID should be that of the arena. You may also want to adjust the SESSION_TIME depending on how long you want the games to last.

To test, publish both places and test on a live server.

Using Persistent Data Stores

A large part of any game, especially a fighting game, is progression. If a player loses all their data—that is, progress—they're going to be less inclined to play again. To address this issue of saving player data, you have an API to load, save, and modify data called Data storeservice. This is perfect for saving player points. Roblox offers persistent data stores for free, which uses a simple structure and API functions to read, save, and alter data. Any and all data is stored in connection to a key, and it can be accessed from any place or server within a game. This can be used to store player data, such as score, to save points in a story and more.

Within our game, this will be incredibly useful for saving player data between sessions. Additionally, if you decide to teleport between places, it will allow you to save and load a players' previous score when they return to the lobby after a game.

Data stores Supports and Limits

Below is some general information about the type of data that can be stored and how often it can be accessed.

▶ Supports strings.

▶ Supports integers and floats.

▶ Key/name/scope length must not exceed 50 characters.

▶ Stored strings must not exceed 65,536 characters.

▶ If the same key is called too often by any number of servers within a game, it may exceed the game limit. Therefore, it's advised to personalize keys to each user so they're not calling the same key.

▶ Write requests require a six-second cooldown.

▶ Request per minute limits:

　▶ Get (for example, `GetAsync()`) = 60 + numPlayers × 10

　▶ Set (for example, `SetAsync()`) = 60 + numPlayers × 10

　▶ Get Sorted (for example, `GetSorterAsync()`) = 5 + numPlayers × 2

　▶ On Update (for example, `OnUpdate()`) = 30 + numPlayers × 5

Data stores keep information in dictionary form using key-value pairs. For example, in Table 17.2, the player's unique UserId is being used as a key, and the value is the player's numeric score. It's recommended that you use UserIds rather than PlayerNames because PlayerNames may change.

TABLE 17.2　Basic Player Data Store

Key (UserId)	Value (Player's Score)
000001	20
000002	62

NOTE

`GetGlobalDataStore()` **and** `GetDataStore()`

`GetGlobalDataStore()` is equivalent to `GetDataStore()` as both are global.

▼ TRY IT YOURSELF

Saving Fighting Game Data

Now that you have the game set up and going, add a layer of progression to the game so that players have a reason to come back. To accomplish this, you can add a point to the player's score every time they kill a player and save the results. To foster competition, you can expand this to a public leaderboard with the top 10 players.

1. To create a leaderboard, add a new script named LeaderBoard in ServerScriptService with the following code:

```
local Players = game:GetService("Players")
local ServerStorage = game:GetService("ServerStorage")

local PlayerPointUpdater = ServerStorage.PlayerPointUpdater

local function onPlayerAdded(player)
      local leaderstats = Instance.new("Folder")
      leaderstats.Name = "leaderstats"
      local score = Instance.new("IntValue")
      score.Name = "Score"
      score.Parent = leaderstats
      leaderstats.Parent = player
end

local function addScore(player)
      local leaderstats = player:WaitForChild("leaderstats")
      local score = leaderstats.Score
      score.Value += 1
end

Players.PlayerAdded:Connect(onPlayerAdded)
PlayerPointUpdater.Event:Connect(addScore)
```

2. In ServerStorage, add a BindableEvent object. Rename it **PlayerPointUpdater**. This will be used in the next section to signal for a change in score.

3. In Game Settings, Security, turn on Enable Studio Access to API Services, which allows datastore usage.

4. Copy the following code into a new script within ServerScriptService and name it **PlayerData**. This will use GetAsync() to load the player's saved data and SetAsync() to update the player's saved data.

```
local Data storeservice = game:GetService("Data storeservice")
local Players = game:GetService("Players")
local ServerStorage = game:GetService("ServerStorage")
```

```
local PlayerPointUpdater = ServerStorage.PlayerPointUpdater
local LeaderboardScore = Data storeservice:GetDataStore("LeaderboardScore")

local function LoadData(player)
        local key = "Player_" .. player.UserId
        local score = player:WaitForChild("leaderstats").Score
        local success, data = pcall(function()
                return LeaderboardScore:GetAsync(key)
        end)

        if success then
                score.Value = data
        else
                score.Value = 0
        end
end

local function SaveData(player)
        local key = "Player_" .. player.UserId
        local score = player:WaitForChild("leaderstats").Score
        local success, data = pcall(function()
                return LeaderboardScore:SetAsync(key, score.Value)
        end)
end

Players.PlayerAdded:Connect(LoadData)
Players.PlayerRemoving:Connect(SaveData)
```

This script loads all player data when they join. If it fails to load, the related leaderboard property is deleted, and any subsequent data isn't loaded or saved to avoid any data loss.

Now go back to your sword script from earlier and modify it to fire PlayerPointUpdater when the player kills someone.

Copy the highlighted lines into your SwordController script:

```
-- Top of script not shown
local Players = game:GetService("Players")
local ServerStorage = game:GetService("ServerStorage")

local PlayerPointUpdater = ServerStorage.PlayerPointUpdater

local tool = script.Parent
local swordBlade = tool.Handle --Change to match your sword blade
local humanoid, animation, player

local canDamage = true
```

```
local isAttacking = false

local function onEquipped()
-- Full function not shown
end

local function awardPoints(otherHumanoid)
    -- Check if other player has no health. If so, award points
    if otherHumanoid.Health <= 0 then
        PlayerPointUpdater:Fire(player)
    end
end

local function onDetectHit(otherPart)
    local partParent = otherPart.Parent
    local otherHumanoid = partParent:FindFirstChildWhichIsA("Humanoid")
    -- Make sure blade doesn't harm controlling player
    if otherHumanoid and otherHumanoid == humanoid then
        return
    -- Checks that there's another player and attack isn't on cooldown
    elseif otherHumanoid and isAttacking and canDamage then
        canDamage = false
        otherHumanoid:TakeDamage(DAMAGE)
        awardPoints(otherHumanoid)
    end
end
-- Bottom of script not shown
```

At this point, you should have a bare-bones game in which players are teleported to an arena where they are given a weapon and can battle each other for points. This code can be expanded on to create a game loop where points reset between rounds, but players are still able to see their total number of kills and deaths between rounds. What other types of mechanics can you think of to include in the game loop?

Data Store Functions

The previous code used GetAsync() and SetAsync() to track player data, but Roblox has a wide range of ways of accessing and modifying data. Here we explain how and why you would use them:

▶ **GetAsync()** allows you to fetch data from a datastore when a key is specified so you know which data you're trying to grab. You use keys for identifying and labeling different types of data, much like with variables.

```
local Score = LeaderboardScore:GetAsync(userId)
print(Score)
```

▶ `SetAsync()` allows you to set new data whether that be with a new key or overwriting old data with an existing one. It requires both the key and data to be set as parameters.

```
LeaderboardScore:SetAsync(userId,10)
```

▶ `UpdateAsync()` works differently than other datastore functions because it takes the key and a function with the update logic as parameters. It attempts to save the data as many times as necessary without any additional logic and should be used if the previous data in the key was important or the data may have been accessed at the same time on another server to reduce data corruption.

```
local updatedScore = LeaderboardScore:UpdateAsync(userId, function(oldScore)
        local newScore = oldScore + 1
        return newScore
end)
print("UpdateAsync: "..updatedScore)
```

▶ `IncrementAsync()` is a way to do the same as the `UpdateAsync()` code in a lot fewer lines. IncrementAsync increments a saved integer by the specified amount, like so:

```
local score = LeaderboardScore:IncrementAsync(userId,1)
print("IncrementAsync: "..score)
```

▶ `RemoveAsync()` offers a simpler way of performing an action, which could be done using an alternative function. RemoveAsync allows you to remove the data associated with a specific key.

```
local removedScore = LeaderboardScore:RemoveAsync(userId)
print("RemoveAsync: "..removedScore)
```

`UpdateAsync()` Versus `SetAsync()`

Although both functions are extremely powerful, it's often advised to use `UpdateAsync()` over `SetAsync()` for updating data in many cases. These include

▶ Updating an existing value not changing or creating new data—for example, `data = oldData + 50`.

▶ Data may be being changed in multiple servers simultaneously or within a short time span.

▶ If the preceding case occurs. It will call `UpdateAsync` again to make sure no data is mistakenly overwritten.

Alternatively, you may prefer to use SetAsync if one of the following is the case:

▶ Creating a new key with data or the new data holds no relation to the last.

▶ The possibility of the data being edited in multiple servers in a short time span does not exist.

Protecting and Responding to Errors

Protecting player data and responding to errors is critical to making successful data. As such, you use something called a pcall to save code if it errors. It can error for a multitude of reasons, including Data storeservice being down or calling the same function too often. Here's an example using pcall:

```
local success, data = pcall(function()

end)
```

What Is a pcall?

A pcall is a special Lua global protected function that acts as a shield from the rest of the script. Any code within it is run in "protected mode" and runs the code, error or not. If the code does error, the pcall catches it and returns a status code (bool) as well as some information about the issue.

We can expand on the first example to incorporate our datastore, protecting our :GetAsync() in the pcall, and then return the fetched datastore data unless it fails, in which case "data" is overwritten by the error info. If this happens, success equals false and our script prints the error instead:

```
local success,data = pcall(function()
return LeaderboardScore:GetAsync(userId)
end)
if success then
print("Did not error, result: "..data)
else
print("Did error, result: "..data)
end
```

Protecting Against Data Loss

Data loss is and can be extremely frustrating both for you and your players, which is why building proper defenses to reduce the risk of data loss is incredibly important. These defenses include the following:

▶ Using UpdateAsync instead of SetAsync.

▶ If loading player data fails, do not attempt to save it.

▶ If something fails, notify the player about the issue.

▶ Save data on an event basis (that is, player gets here, player does x).

Summary

In this hour, you learned how to build a basic fighting game. You learned about the Roblox tool container, teleporting within and between servers, as well as saving your score data between sessions.

Q&A

Q. How do you display an image on a tool in the toolbar?

A. Tool.TextureId.

Q. Do tools require models or parts to function?

A. No, tools can work with or without any 3D objects—for example, spells may have only an icon.

Q. What part of a tool welds to the player's right hand?

A. The handle.

Q. What service is used to teleport players between places?

A. TeleportService.

Q. Describe the purpose of RemoveAsync.

A. RemoveAsync removes a data entry.

Workshop

Now that you have finished, review what you've learned. Take a moment to answer the following questions.

Quiz

1. When does tool.Activated fire?

2. What character property do you use to teleport players within a place?

3. What do CFrames control?

4. How do you teleport a group of people between places?

5. Suggest two ways to protect against data loss.

6. What's the purpose of pcall?

Answers

1. tool.Activated fires when the player clicks while the tool is activated.

2. HumanoidRootPart is the character property used to teleport players within a place.

3. CFrames control position and rotation/orientation.

4. You use TeleportPartyAsync() to teleport a group of people between places.

5. Using UpdateAsync() and saving regularly are two ways to prevent data loss.

6. To run code in protected mode, errors don't stop within or affect the code without, instead catching the error and returning a status code.

Exercises

The exercises combine a number of different things you've learned in this hour. If you get stuck, don't forget to refer to the previous pages in this hour.

There are always ways in which you can improve your game, one of which is by creating more competition. Your task for the first exercise is to make a recording of the top global player. Every new score entry should be checked against it. If a player surpasses the top score, this new value and player is saved on a unique datastore.

For the second exercise, create a new script and data store that assigns a player points every time they join. One condition, though: You are not allowed to use `GetAsync` or `SetAsync`. These points should be displayed in the output window or onto the leaderboard.

HOUR 18
Multiplayer Code and the Client-Server Model

What You'll Learn in This Hour:

▶ What the client-server model is

▶ What `RemoteEvents` and `RemoteFunctions` are

▶ How to use `RemoteEvents` and `RemoteFunctions` in your game

▶ What server-side validation is

▶ How to use teams

▶ How to set network ownership

Roblox is all about playing with friends in massive multiplayer experiences. Designing your game to work with multiple players is essential if you want an interactive, social game. To do this, you'll need to understand that the basics of code can run locally on the player's device or on Roblox's servers and what that means for both the performance and security of your game.

In this hour, you learn the basics of the client-server model. You also learn how to design for a multiplayer experience with teams of players and gain an understanding of how to use `RemoteEvents` and `RemoteFunctions` to send data between clients (players) and the server.

The Client-Server Model

Games on Roblox, along with many other multiplayer games, use a type of network structure known as the client-server model (Figure 18.1). When you connect to a game, regardless of the device you're playing on, you are known as a client. You are connected to a server that hosts an instance of the game you are playing. In the next section, you'll learn more about the two types of scripts in Roblox games, along with replication and its importance in the client-server model.

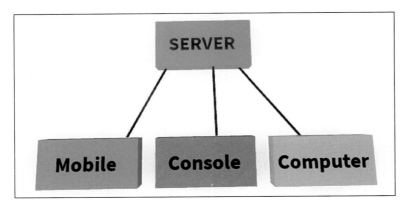

FIGURE 18.1
Diagram showing how clients connect to a server.

Scripts and LocalScripts

There are two types of scripts: a Script, which runs on the server, and a LocalScript, which runs on the client (Figure 18.2). By using these scripts with events (such as RemoteEvent, described later in this hour), the client and server are able to communicate with each other. This is essential for any interactions in your game to occur—for example, pressing a button and changing the position of a part in your game.

FIGURE 18.2
The two types of scripts in Roblox.

Replication

In Roblox, most changes you make on the client are not replicated to other players automatically. This means that if you change the position of a part using a LocalScript, only you will see the change. Other players, and the server, will not be able to see this change.

This is very important to remember because it can be useful in many scenarios. For example, if you want to display something to a player during the tutorial in your game, you could display this item using a LocalScript on the client, and only that player would see it. However, if you want to make changes that every player can see, you need to send a request to the server by using a RemoteEvent or RemoteFunction, which are explained in the next section.

Automatic Replication

Some changes on the client *do* automatically replicate to the server and other players.

These include animations, sound, ClickDetectors, humanoid changes (for example, sitting or jumping), and part physics.

What Are `RemoteFunctions` and `RemoteEvents?`

`RemoteFunctions` and `RemoteEvents` are objects you insert into the game to talk to the server. Whenever a client wants to request something, or talk to the server, a message is fired to the server using one of the objects shown in Figure 18.3.

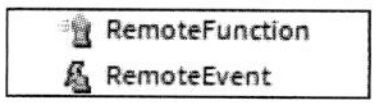

FIGURE 18.3
A `RemoteFunction` and `RemoteEvent` instance.

`RemoteEvents` are used for one-way communication. A `RemoteEvent` can be fired from a client to the server, or the server can fire to an individual client or all clients in the game. The three options for `RemoveEvents` are `FireServer()`, `FireClient()`, and `FireAllClients()`. For example, using `FireServer()` would look like:

```
RemoteEvent:FireServer()
```

Inside the `FireServer` function, data such as a string can be added to pass information to the server, as shown in the following example. There are many things you can pass to the server, such as a CFrame position, `Color3` RGB value, or even a table of data.

```
RemoteEvent:FireServer("Hello")
```

The server receives the message and carries out any action necessary.

On the other hand, a `RemoteFunction` is used for two-way communication. Once an event has been fired, or invoked, the receiver responds with a reply. The two `RemoteFunction` options are `InvokeServer()` and `InvokeClient()`. For example, if a client sends a message to the server like so:

```
local reply = RemoteFunction:InvokeServer("Hello")
```

the server would receive this and reply with

```
local function anyfunction(player)
    return 5
end


RemoteFunction.OnServerInvoke(anyfunction)
-- When the RemoteFunction is invoked by the server, run the function called
    anyfunction
```

The value of 5 would then be sent back to the client. If the client printed the value of `reply`, the output would be 5. This is useful for returning positions, tables of data, or even models in your game. An example would be a `RemoteFunction` that clones a model on the server and returns it for use in an item placer.

NOTE

Minimize

You should use `InvokeClient()` as little as possible. The server will be waiting for a reply from the client. If the client is laggy, it may be waiting a long time! If the client disconnects/leaves the game, the function will error. When you do use an `InvokeClient`, you should wrap the function in a pcall, as mentioned in Hour 17.

How to Use `RemoteEvents` and `RemoteFunctions`

One important thing to remember about these two instances is that they must be in a location where both the server and client can access them. A good place is `ReplicatedStorage`, but you could also put them in the Workspace. Consider storing them in a folder in `Replicated-Storage`, as shown in Figure 18.4.

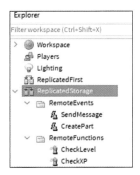

FIGURE 18.4
Storing `RemoteEvents` and `RemoteFunctions` in folders in `ReplicatedStorage`.

Creating a `RemoteEvent`

Begin by creating a `RemoteEvent` so you can try communicating between the client and server. First, click `ReplicatedStorage` and insert a new `RemoteEvent` instance, as shown in Figure 18.5.

FIGURE 18.5
Insert a new `RemoteEvent`.

You're going to create an event that asks the server to create a new part in the Workspace. For now, rename it to `SendMessage` (Figure 18.6).

FIGURE 18.6
Rename the event.

Next, create a LocalScript and put it in StarterPlayerScripts (Figure 18.7).

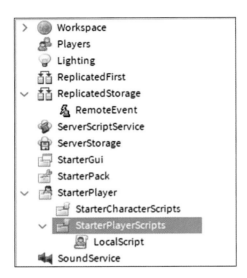

FIGURE 18.7
Insert a LocalScript into StarterPlayerScripts.

Now write the code to fire an event. Inside the LocalScript, write the following:

```
local remoteEvent = game:GetService("ReplicatedStorage"):WaitForChild("SendMessage")
remoteEvent:FireServer()
-- Fire an event to the server
```

You now have a way of sending a message to the server from a client, and you need to create the script that will respond to that message on the server. Create a script and put it in Server-ScriptService (Figure 18.8).

Inside that script, write code to listen for the SendMessage event and create a new part when it receives a message from any client:

```
local remoteEvent = game:GetService("ReplicatedStorage"):WaitForChild("SendMessage")

local function createPart(player)
    -- Create a part and rename it to the name of the player
    local part = Instance.new("Part")
    part.Parent = game.Workspace
    part.Name = player.Name
end

remoteEvent.OnServerEvent:Connect(createPart)
-- When the remoteEvent is triggered by the server, run the function called
      createPart
```

Every time a client fires this event, the server creates a part and changes the name of it to the name of the client (player) who fired it. Go ahead and test it!

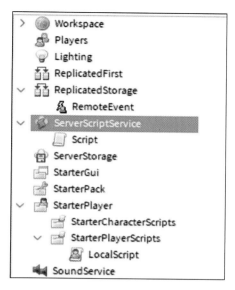

FIGURE 18.8
Insert a script into ServerScriptService.

Experiment with `RemoteEvent`s and `RemoteFunction`s!

You can change the function to do whatever you'd like! For example, changing the color of a part, sending a message to another player, deleting a part, or even triggering an explosion! Try experimenting with `RemoteFunction`s too; get the server to reply with a value to the client or return an object. It is important you use server-side validation, however, which we discuss in the next section.

Server-Side Validation

Imagine you have a `RemoteEvent` called "PurchaseItem," and you pass in two parameters: the name and price of the item.

```
remoteEvent:FireServer("PetDog", 100)
```

This is a *very* bad idea. Any guesses why?

A hacker, or exploiter, could fire the event but change the price to 0 and get an item in the game for free! This is where something called *server-side validation* is very important. This is where the server checks, or validates, a value passed from the client. For example, the server might store Value objects in `ServerStorage`. The NumberValue objects shown in Figure 18.9 hold numerical data and are a convenient way to store the prices for a shop.

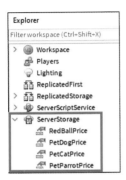

FIGURE 18.9
NumberValue objects in ServerStorage, which holds the prices of items in the game.

When the server receives the event, it checks the price of the Pet Dog and sees that it is in fact 100 Cash, not 0 like the exploiter said it was. The exploiter would be charged 100 or denied the purchase. Exploiters often try to change the amount of money they have or the level they are, but these changes occur on the client. If you always validate on the server, the changes won't cause an issue!

Teams

So, you've made a competitive, multiplayer game. The only problem is, everyone is playing together, and there's no competition! Sounds like you need to create teams. In this section, you learn how to add teams to your game and assign players to a team.

Adding Teams

First, open the Model tab and click the Service button in the Advanced section to open the Insert Service dialog box (Figure 18.10).

FIGURE 18.10
Open the Model tab, find the Advanced section, and click the Service button.

Select `Teams` and click Insert (Figure 18.11). This creates a special folder to hold all your teams.

FIGURE 18.11
Use Insert Service to add the Teams folder.

Select the Teams folder and insert a new team (Figure 18.12). You can change the name and color of this team in the properties.

FIGURE 18.12
Insert a team.

NOTE

Nil

Setting `Player.Neutral` to true will change the team to nil. You could check this by printing `game.Players.LocalPlayer.Neutral` in a LocalScript.

Auto-Assigning Players to a Team

In the properties, you can select the check box for Auto-Assignable (Figure 18.13) if you want players to automatically join this team when they join the game. For example, you could set this if you wanted a Spectator or Lobby team.

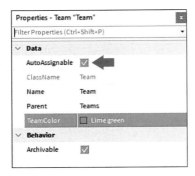

FIGURE 18.13
Select AutoAssignable in the Properties tab to automatically assign players to a team.

Manually Assigning Players to a Team

This example shows you how to put every player in the game onto a team called WinningTeam. To do this, loop through all the players and change their Team value like this:

```
-- Loop through every player in the game and change their team
for _, player in pairs(game.Players:GetChildren()) do
    player.Team = game.Teams.WinningTeam
end
```

You could also use the TeamService like so:

```
-- TeamService
local Teams = game:GetService("Teams")

-- Loop through every player in the game and change their team
for _, player in pairs(game.Players:GetChildren()) do
    player.Team = Teams["WinningTeam"]
end
```

This method is often more useful if you create or delete teams during gameplay. GetTeams() creates a table of all the current teams in the game. Using it can also make randomizing teams easier, like so:

```
-- TeamService
-- GetTeams() returns a table of all current teams in the game
```

```
local Teams = game:GetService("Teams"):GetTeams()

-- Loop through every player in the game and change their team
for _, player in pairs(game.Players:GetChildren()) do
    player.Team = Teams[math.random(1, #Teams)]
        -- Picks a random team out of the available teams
end
```

TRY IT YOURSELF ▼

Experiment with Other Ways to Assign Players!

There are more complex ways to assign players to teams—for example, grouping friends using the `isFriendsWith()` method and moving that group to a specific team is one. That's beyond the scope of this book, but you can experiment with it and have a go yourself!

Network Ownership

Whenever an object (for example, a vehicle or part) in a Roblox game moves around, its physics are calculated by Roblox. So that the server doesn't have to calculate physics for every object in the game, the server lets some of the clients calculate physics for the objects close to them. The client that is calculating the physics of an object is known as the "owner," hence the name "network ownership."

Most of the time, network ownership works fine. However, any physics updates managed by the client have to be sent to the server. For example, if a part is flying through the air and flies past multiple players, each player will have their turn at being the owner. This can cause visible lag as the part changes network ownership. To make sure the part moves smoothly, you can assign one particular network owner manually.

You can set the network ownership of a part (on the server) like this:

```
game.Workspace.Part:SetNetworkOwner(player)
```

You can also print `GetNetworkOwner()` to view the current network owner of a part.

NOTE

Network Ownership and Anchored Parts

Network ownership cannot be set for an anchored part. A warning will be printed in the output window if you try to do this!

Summary

In this hour, you've learned the basics of the client-server model and how it's used in Roblox to allow communication between clients and the server. You've learned what `RemoteEvents` and `RemoteFunctions` are and how to use them in your game. These tools are two of the most important things you'll need in your game development journey! We've explained what server validation is and why it's so important. Finally, you've learned how to create teams to make fun multiplayer games, and you've been introduced to network ownership and how you can use it.

Q&A

Q. Do fired `RemoteEvents`/`RemoteFunctions` execute in order?

A. Yes, even if they don't arrive in order, they will be executed in the correct order.

Q. Will my script wait for a `RemoteFunction` reply?

A. Yes, `RemoteFunctions` "yield" the script until a reply is received. This is why invoking the client can be a bad idea! `RemoteEvents` do *not* yield, however.

Workshop

Now that you have finished, take a few moments to review and see if you can answer the following questions.

Quiz

1. True or False: Every change made on the client replicates to the server.

2. True or False: There are two types of scripts.

3. To protect `RemoteEvent` or `RemoteFunction`, you should always use server-side _____.

4. True or False: `RemoteEvents` are for one-way communication.

5. True or False: Setting `Player.Neutral` doesn't change the player's team.

6. What does `InvokeClient()` do?

7. Which networking model does Roblox use for game servers?

8. Network ownership determines who calculates the _____ of a part.

Answers

1. False. Only some changes (like humanoid or sound) replicate.

2. True. **LocalScripts** run on the client and **Scripts** run on the server.

3. Validation.

4. True. `RemoteEvent`s are one way—`RemoteFunction`s expect a reply!

5. False. Their team will be set to nil.

6. `InvokeClient()` is used with a `RemoteFunction` to send a message to the client. The client receives this and replies.

7. The client-server model.

8. Physics.

Exercises

This exercise combines a number of different things you've learned this hour. If you get stuck, don't forget to refer to the previous pages in this hour. Try to make a GUI button which, when clicked, sends a `RemoteEvent` message to change the player's team. The `FireServer()` event should contain the team name the player wants to change to. Don't forget to validate on the server that the team exists!

1. Create a ScreenGui, and insert a TextButton.

2. Inside the TextButton, add a LocalScript.

3. Create a `RemoteEvent` in `ReplicatedStorage`, and rename it to a suitable name.

4. Insert a script into `ServerScriptService` that runs when the `RemoteEvent` you created has been fired by a player. Once it has been fired, change the team of the player who fired it. Don't forget to validate the team name!

5. Go back to the LocalScript and fire the `RemoteEvent` when the player clicks the button. You can use the `MouseButtonClick1` event to detect the TextButton being cleared. Don't forget to use `FireServer()` with a team name argument!

6. Within the script, use the two functions you've created to transition to the other animation. (So, when the MoveLeft animation is finished, start MoveRight, and vice versa.)

7. Go ahead and test it!

8. **Bonus:** Create a TextBox that enables the player to type the name of the team they want to join. Send that as an argument when you `FireServer()`. Again, don't forget to validate that the team exists on the server!

Bonus Exercise: Create a laser part that fires from the player toward another part or position. Use network ownership by setting the network owner of the laser part.

1. Create a new part and parent it to the workspace.

2. Change the Color3 and Material if you want.

3. Set the CFrame of the part. You can use `player.Character.Head.Position` as an example.

4. Set the velocity of the part to make it move! Velocity uses Vector3.

5. Set the network owner. You can use `SetNetworkOwner(nil)` to let the server take ownership or `SetNetworkOwner(player)`.

6. Get the direction the player is facing by using `local direction = player.Character.Head.CFrame.lookVector`. You could add this on to the Velocity value like so: `part.Velocity = direction * 20 + Vector3.new(0,20,0)`.

7. **Extra:** Test with different network owners (on a live server, not Studio) to see if you can spot the difference!

HOUR 19
Module Scripts

What You'll Learn in This Hour:

▶ What a module script is

▶ How module scripts on the client and server side differ

▶ How to use module scripts to store information

▶ How to code a game loop using module scripts

When coding games, there are lots of situations where you may need to use a chunk of code in several different places or make it accessible by a number of scripts. For example, you may want to give a player items at the end of a quest or when they open a chest.

You could copy and paste code into 20 different treasure chests, but updating that would be a pain. A better way to organize your code—and avoid repeating yourself—is to use a module script. In this hour, you find out what module scripts are, when to use them, and how to put them into practice by creating a game loop.

Getting to Know the Module Script

A module script is a special type of script that exhibits special behaviors that allow functions and variables within it to be referenced and used across multiple scripts. The module script can be used by both the client and the server if it is in an area where both sides can access it.

To begin, create a module script inside ServerStorage, as shown in Figure 19.1.

FIGURE 19.1
A module script located inside ServerStorage.

This is the most common area for storing module scripts because scripts that are inside Server-Storage are not executed when the game is run. However, you also can place module scripts inside ReplicatedStorage, which is particularly useful if you want to house variables and functions that both the client and the server can use.

Understanding the Anatomy of a Module Script

Unlike other script types, a module script does not automatically include `print("Hello world!")`. Instead, the generated code creates a local table, which is then returned (or sent) to the calling script. Any code in the module table, such as values, functions, other tables, and pieces, will be sent within the module table.

```
local module = {}   -- Local table

return module -- Returns the table to where the script was called
```

The first thing you should do with any module script is to rename the script, and then update the table to match, as in the following example:

1. Rename the script with a name that describes what the code is for. This example will be TreasureManager (Figure 19.2).

FIGURE 19.2
The TreasureManager script.

2. Rename the table to match:

```
local TreasureManager = {}

return TreasureManager
```

3. If the variables and functions only need to be used by the module script, set them using the keyword `local` as you normally would.

4. Add a local value named `goldToGive`:

```
local TreasureManager = {}
local goldToGive = 500 -- local variable only usable by the Module Script

return TreasureManager
```

Adding Code That Can Be Used Anywhere

To make the code accessible from other scripts, instead of simply declaring variables and functions as you normally would, you must make sure to add it into the table so that it can be used by other scripts. To do that, reference the table using `ModuleName.InsertNameHere`.

Variable example

```
ModuleName.VariableName  =  100
```

Function example

```
function ModuleName.FunctionName()
end
```

To continue with the example you started earlier, now you want to add a function that can be called from any treasure chest in the game by adding a function named `giveGold` to the table. Do not make it local:

```
local TreasureManager = {}
local goldToGive = 500

function TreasureManager.giveGold() -- Can be called from anywhere
      print(goldToGive .. "gold was added to inventory" )
end

return TreasureManager
```

Using a Module Script

Now that you have your module script filled, it's time to use it. To access a module script from a LocalScript or a regular script, you need to use `require()`, which accepts one argument: the module script's location in the Explorer. You also assign a variable to the result of the function (recall that a module script returns a table), as shown in the following code:

```
-- Getting the module script
local ServerStorage = game:GetService("ServerStorage")
local ModuleExample = require(ServerStorage.ModuleExampleMyModule)

To use the functions and variables inside of the module, use dot notation.
-- Function example
ModuleExample.exampleFunction()
```

You still need to have some code in the treasure chest to call the module script. Keeping that code as light as possible makes it less likely that you have to update it later and makes your code more resistant to hacking. Call `giveGold()` whenever a player touches a treasure chest. Use the following steps:

1. Use a simple part or a model to act as your treasure chest. Insert a script. If you use a model, the script must go into a part.

2. Use `require()` to get the TreasureManager module script.

3. Add code that calls `giveGold()` when the player touches the treasure chest:

```
local ServerStorage = game:GetService("ServerStorage")
local TreasureManager = require(ServerStorage.TreasureManager) -- Gets module
    script

local treasureChest = script.Parent

-- Gives gold when touched by a player
local function onPartTouch(otherPart)
        local partParent = otherPart.Parent
        local humanoid = partParent:FindFirstChildWhichIsA("Humanoid")
        if humanoid then
                TreasureManager.giveGold() --Calls the function from the module
                        script
        end
end

treasureChest.Touched:Connect(onPartTouch)
```

▼ TRY IT YOURSELF

Connect to Leaderboard

See if instead of just running a print statement, you can trigger a change on a leaderboard when the player picks up gold. Refer to Hour 12 for how to use leaderboards.

Studio may automatically suggest the variable or function after the dot, but other times you may need to manually type it. In that case, make sure that it's spelled exactly the same as it is in the module script.

NOTE

Using :WaitForChild() Instead of Dot Notation

Using dot notation may cause errors if the script attempts to require the module script before the module script is loaded in. However, this is unnecessary if your script is in *ServerScriptService* or *ServerStorage*. (Note that the preceding script is in the former.)

Creating a Module Script

You can add functions to do whatever you want, such as creating/destroying parts, manipulating the character, or simply printing something out. Note that *where* you require the module script will affect *who* can see the effects. The next section covers this.

Understanding Client-Side Versus Server-Side Module Scripts

Although module scripts can be accessed and used by both local and server scripts, what it can do depends on which side of the client-server model it is being executed on. Hour 18, "Multiplayer Code and the Client-Server Model," explained the client-server model, as well as what is and isn't replicated when code is run on the client side. Similarly, when a module script is run from a script, its interaction is dependent on what the script running it can interact with. For example, a LocalScript that runs a `ModuleScript` function will *not* be able to access ServerStorage, ServerScriptService, or other areas that are only visible to the server. However, running the same `ModuleScript` function within a regular script will. This allows module scripts to be used on both sides of the client-server model without a potential loophole for exploiters to use. Let's go through an example in which you run a script from both the server and the client to demonstrate:

1. In ServerStorage, add a `NumberValue` object named Secret (Figure 19.3). Set the value to any number you like. ServerStorage can be accessed by only the server.

FIGURE 19.3
Place an object inside ServerStorage.

2. Place a `ModuleScript` inside ReplicatedStorage (Figure 19.4), a container that both the client and server can access.

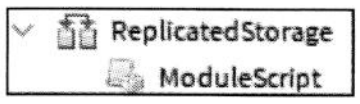

FIGURE 19.4
ModuleScript inside ReplicatedStorage.

3. Name the module script ReplicatedModuleScript.

4. Rename the table to match the name of the module script:

```
local ReplicatedModuleScript = {}
return ReplicatedModuleScript
```

5. Add Secret to your `ModuleScript` function using a variable:

```
local ReplicatedModuleScript = {}
ReplicatedModuleScript.Secret = game.ServerStorage.Secret
return ReplicatedModuleScript
```

```
local ReplicatedModuleScript = {}
ReplicatedModuleScript.Secret = game.ServerStorage.Secret
return ReplicatedModuleScript
```

For the LocalScript and regular script, you use the same code because they can both access the module script located within ReplicatedStorage. Then you attempt to print the value of Secret after obtaining the table:

1. Place a script in ServerScriptService and a LocalScript in StarterPlayerScript.

2. Copy the following code into both scripts:

```
-- Get the required module script
local  ReplicatedModuleScript = require(game.ReplicatedStorage:WaitForChild
      ("ReplicatedModuleScript"))

-- Get and print the variable value
print(ReplicatedModuleScript.Secret.Value)
```

Once you are set up, you can playtest and observe the results. You should see both the printed value from the client side and an error message from the server side. In Figure 19.5, the server (the green line on the top) outputs the SecretKey's value, whereas the client (the blue line on the bottom) produces an error.

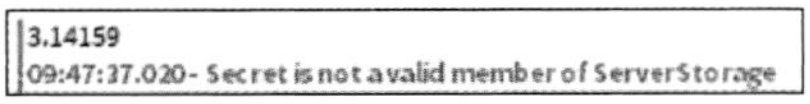

```
3.14159
09:47:37.020- Secret is not a valid member of ServerStorage
```

FIGURE 19.5
The resulting output.

As you can see, while the server had no issue accessing and printing the Secret value, the client just couldn't find it, and instead displays `Secret is not a valid member of Server-Storage`. This separation of functionality in module scripts is particularly useful when dealing with one that can be accessed by both ends of the client-server model. You wouldn't want an exploiter to see what's in your private ServerStorage, would you?

Using Module Scripts: Game Loop

Now that you are a bit more familiar with module scripts, how to use them, and how they differ in functionality between the client and server, you can make something with it: a game loop. A game loop is the cycle a player goes through each time they play your game. In multiplayer games, you might have a round-based game loop where players compete in some fashion to win a match. For this type of game loop, you need to account for three major states (Figure 19.6):

- ▶ **Intermission:** Waiting for players to join or the round to start

- ▶ **Competition:** Where the action occurs

- ▶ **Cleanup:** Where everything is reset to its original state

FIGURE 19.6
The three main components of a round-based game loop.

Establishing Control with Settings

First, create an object that other scripts can reference when they need information regarding each stage:

1. Create a module script inside ServerStorage. You can name it anything you want, although you preferably want something memorable like GameSettings or GameInformation (Figure 19.7).

FIGURE 19.7
Create a module script inside ServerStorage.

2. At the top of the module script, you need to set up the variables that will control how long each part of the game loop lasts. Before doing so, ask yourself these questions:

   - ▶ How long should the intermission last?

   - ▶ How long should each match/round last?

   - ▶ How many players do I need minimum for the game to function properly?

   - ▶ How much time should I have in between each "stage"?

3. Add each as a variable inside the module script.

In the following code, we have filled out a module script with answers to the preceding questions and gave them appropriate names. All times are measured in seconds:

```
local GameSettings = {}

GameSettings.IntermissionTime = 5
GameSettings.RoundTime = 30
GameSettings.MinimumPlayers = 2
GameSettings.TransitionTime = 3

return GameSettings
```

Creating Reusable Round Functions

Next, work on the script that manages the matches.

1. Create another module script inside ServerStorage named RoundManager (Figure 19.8).

FIGURE 19.8
RoundManager module script added in ServerStorage.

2. Inside the RoundManager module script, add the following variables and functions that will be used to send the players to the match and reset everything when the match is over. Right now, print statements are being used as placeholders and functionality can be added later.

```
local RoundManager = {}

local ServerStorage = game:GetService("ServerStorage")
local GameSettings = require(ServerStorage:WaitForChild("GameSettings"))
local RoundManager = {}

function RoundManager.PreparePlayers()
    print("The match is beginning...")
    wait(GameSettings.TransitionTime)
end

function RoundManager.Cleanup()
    print("The match is over. Cleaning up...")
    wait(GameSettings.TransitionTime)
end

return RoundManager
```

Creating the Main Engine: The Game Loop

Now it's time to work on the main engine of the system, which is the loop itself:

1. Create a regular script, this time inside ServerScriptService (Figure 19.9) so that it runs as soon as the game begins.

FIGURE 19.9
A script (renamed to GameLoop) located inside ServerScriptService.

2. Set up the environment by getting everything you need: Get both module scripts from ServerStorage, as well as some other Roblox services such as Players and RunService.

   Here is the initial setup of the main loop script. Notice `WaitForChild()` inside the `require()` calls. This is to make sure the module scripts are loaded before the rest of the code in this script runs.

```
-- Services
local RunService = game:GetService("RunService")
local ServerStorage = game:GetService("ServerStorage")
local Players = game:GetService("Players")

-- Module Scripts
local GameSettings = require(ServerStorage:WaitForChild("GameSettings"))
local RoundManager = require(ServerStorage:WaitForChild("RoundManager"))
Create the infinite loop using while true do:
-- Main Loop
while true do
    -- Code inside will repeat every frame
end
```

3. Check to make sure that you have enough players before you attempt to begin a match.

   You can do that using an `if` statement that compares the number of players to the minimum number of players required to start.

   Note the # symbol, which returns the length of a table. In this case, the # will get the length of the player list, obtained through `Players:GetPlayers()`:

```
-- Main Loop
while true do
-- Any code in here will be repeated every frame
    if #Players:GetPlayers() < GameSettings.MinimumPlayers then
        wait()
    end
end
```

4. Once there are enough players, you begin the round and wait for it to end:

```
-- Main Loop
while true do
-- Any code in here will be repeated every frame
    if #Players:GetPlayers() < GameSettings.MinimumPlayers then
        wait()
    end

    wait(GameSettings.IntermissionTime)

    RoundManager.PreparePlayers()

    wait(GameSettings.RoundTime)

    RoundManager.Cleanup()
end
```

And with that, you have the foundation for your game loop. Feel free to test it out on a live server with friends or use the testing tool with more players than minimally required in the Test tab at the top (Figure 19.10).

FIGURE 19.10
Use the testing tool to start a local server with up to eight players on your own machine.

▼ TRY IT YOURSELF

Completing the Game Loop

With the foundations in place, you can begin messing around with what the round manager does to prepare players and clean up. Feel free to build a lobby that players can mingle in before the match begins, give players weapons before, or anything in between! You can also improve the game loop by using BindableEvents to signal the start and end of a round (so that a player doesn't need to wait until time's up if certain conditions are met).

Summary

In this hour, you learned about module scripts and how they can help you organize your code better and avoid repeating functions. Applying your previous knowledge of the client-server model, you learned what module scripts can do depending on which side of the model they are being used on. Finally, you learned how to apply the ideas of module scripts to a system using the game loop example.

Q&A

Q. **Can module scripts require each other?**

A. Yes. Having two module scripts require each other is generally OK, but more than that can lead to issues where they are all waiting for each other to load.

Q. **Where can I place my module scripts?**

A. Best practice is to place them in ServerStorage if they are only used by regular scripts on the server side, and ReplicatedStorage if they are used by both LocalScripts and server scripts.

Workshop

Now that you have finished, take a few moments to review to see whether you can answer the following questions.

Quiz

1. True or False: Module scripts run by themselves.

2. True or False: Module scripts can *only* be placed within ServerStorage.

3. For a Module script to be used, you must use _____.

4. True or False: Local variables or functions are accessible outside of the module script.

5. True or False: Module scripts run on the client side can see everything that the server sees.

6. True or False: Module script code can be used by multiple scripts.

7. A module script returns a _____, which contains a combination of variables and functions.

8. You use module scripts as a way to better _____ your code.

Answers

1. False. Module scripts must be run or accessed by another script (such as a LocalScript or Script).

2. False. Module scripts can be placed anywhere as long as they are accessible by the script trying to use them. (See the Q&A section for best practices.)

3. For a module script to be used, you must use `require`.

4. False. Local variables or functions can only be accessed within the module script itself.

5. False. Module scripts on the client side can only see what the client sees (no peeking at the ServerStorage).

6. True. Module scripts can be used by any number of scripts.

7. A module script returns a table.

8. You use module scripts as a way to better organize your code.

Exercises

This exercise combines things you've learned this hour. If you get stuck, don't forget to refer to the previous pages in this hour. Try to make some bricks, which, when touched, will give you a different amount of currency—but run off the same module script!

1. Create a script inside ServerScriptService that creates a leaderstat of a currency (it can be anything you want).

2. Create several parts, with a script inside that runs when a player touches it.

3. Inside ServerStorage, create a module script and rename it to an appropriate name.

4. Inside the module script, create a function that gives currency based on the part that the player touches. (Hint: Use *if* statements to check.)

5. Call the function inside the script using the `require` keyword and pass the `Character` and `Part` as arguments.

6. **Extra:** Instead of using multiple scripts for the parts, try to use one using a `for` loop!

Bonus Exercise: Create a button that says different things depending on whether you're calling it from the client side or the server side, with one module script and only one function!

1. Create a module script inside ReplicatedStorage.

2. Inside the module script, write a function that prints text depending on whether the server or the client is running the code using the `RunService:IsClient` and `RunService:IsServer` functions (you will need to insert `RunService` by creating a variable and setting the service to it).

3. Create a script inside ServerScriptService.

4. Inside the script, use `require` to access and call the module script function.

5. Create a ScreenGui and a TextButton. Feel free to change the names and the text.

6. Create a LocalScript inside the TextButton, which also uses `require` to access and call the `ModuleScript` function.

7. **Extra:** Use a `RemoteEvent` from the previous chapter so that the LocalScript can call the server side of the module script.

HOUR 20
Coding Camera Movements

What You'll Learn in This Hour:

▶ What cameras are
▶ How to script the camera
▶ How to use the render step
▶ How to rotate and offset the camera

As the eyes of the player, the camera is the unsung hero in how players experience a game. A clunky, unwieldy camera destroys the player experience, whereas a refined system can make the player feel like they are more involved in the action. A camera may zoom in on an NPC as the player talks to them or pan out to give them a better view, all without even being noticed by the player.

This hour introduces you to cameras and how you can animate and rotate them smoothly to create quick camera actions, cinematic camera spins, and a camera shake.

Introduction to Cameras

Cameras affect the mood of a player as they play. When game designers are trying to convey how the player should feel at a certain point in time, they'll often change the focus of the camera. For example, if the designer wants players to feel scared and a bit claustrophobic, they bring the camera in close. Figure 20.1 shows a scene where a designer has used the camera to create a mood.

FIGURE 20.1
Camera tight behind the player to create tension.

If the designer instead wants the player to feel more adventurous and carefree, they set the camera further back to make more of the world visible, as in Figure 20.2.

FIGURE 20.2
Wide shot emphasizing open-world gameplay.

Within Roblox, each player, or *client*, has a local camera object. It's the way in which the player sees the 3D world rendering on the local device, be that phone, tablet, PC, Mac, or Xbox. The default camera object resides in the Workspace (Figure 20.3).

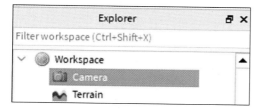

FIGURE 20.3
Camera example.

Camera Properties

Now that you know where the camera is and know how to access it, what exactly can you do with it? First, we'll go through the different camera properties and then jump into some examples. Table 20.1 displays some of the default camera properties that come along with the camera object. In the next section, we offer examples of how you can modify properties to improve the player experience.

You can access properties like so:

```
local currentCamera = workspace.CurrentCamera
currentCamera.FieldOfView = 100
```

TABLE 20.1 Camera Properties

Property	Explanation	Examples
CFrame	The coordinate frame of the camera (position and rotation, as covered in Hour 14, "Coding Animation")	Cutscene Camera tweening Beauty shots
Focus	A CFrame property containing the 3D area to prioritize for rendering, by default, the player's Humanoid	Improve visual fidelity in areas away from the player character, such as cutscenes.
FieldOfView	Also known as FOV, the angular size at which the camera can view the world, typically measured in degrees	Sniper scope—FPS, such as Phantom Forces Binoculars, such as Jailbreak Emotional response, such as creating a claustrophobic or confident feeling

Basic Camera Manipulation

What an individual player sees is by necessity different than the view every other player sees. This affects how you handle camera code. Like any code that affects what an individual player sees, code affecting the camera typically needs to be within a LocalScript, and LocalScripts that deal with cameras should be stored within StarterPlayerScripts (Figure 20.4).

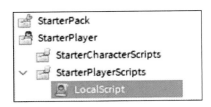

FIGURE 20.4
A LocalScript stored within StarterPlayerScripts that will be used to make changes to the camera.

To manipulate the camera, you first have to change the camera type to one that allow changes:

```
local camera = workspace.CurrentCamera
camera.CameraType = Enum.CameraType.Scriptable
```

You can access properties like so:

```
local camera = workspace.CurrentCamera
camera.CameraType = Enum.CameraType.Scriptable
camera.FieldOfView = 30
```

Coding a Camera Move

In this section, we explain how to create a camera that tweens to a new spot in an environment before returning to normal. This type of motion could be used to show players events that are happening nearby, like a puzzle that was just solved or a door opening. Use the following steps:

1. To mark where you want the camera to move, create a wedge part named EndGoal. Wedges are good because you can easily tell which way they are pointing (Figure 20.5).

2. In StarterPlayerScripts, add a LocalScript and name it (Figure 20.6).

FIGURE 20.5
A wedge that will be used to visualize where the camera will point.

FIGURE 20.6
LocalScript renamed CameraMove.

3. Get the TweenService, the `CurrentCamera` object, and the wedge you just created.

```
local TweenService = game:GetService("TweenService")

local currentCamera = workspace.CurrentCamera -- The camera

local endGoal = workspace.EndGoal -- The wedge
```

4. Add a `wait()` to give this example time to spawn and make the camera scriptable.

```
wait(3) -- Giving time for character to load in this specific example
currentCamera.CameraType = Enum.CameraType.Scriptable
```

5. Set up the tween to move the camera. Refer to Hour 14 if you want to add additional arguments to the tween. This will take the current position of the camera and tween it to match the wedge.

```
-- Set up the camera tween
local tweenInfo = TweenInfo.new(10)

local goal = {}
goal.CFrame = endGoal.CFrame

local CameraAnim = TweenService:Create(currentCamera,tweenInfo,goal)
```

6. Set up a function that will return the camera back to normal when the tween completes.

```
-- Restores normal camera after short pause
local function returnCamera()
    wait(3) -- Give players a chance to look at view
    currentCamera.CameraType = Enum.CameraType.Custom
end

CameraAnim:Play()
CameraAnim.Completed:Connect(returnCamera)
```

Always make sure you return the camera back to normal when you are done; otherwise, the player won't be able to go back to normal.

Using the Render Step

The prior method used tweens, but often you need to move the camera without a preset animation or for a specific amount of time. Rather than using a loop to move the camera, you want to use the *render step*. As the player views a game on their screen, the image shown is very rapidly refreshing to give the illusion of the player smoothly moving about the game environment. The time during which all of the calculations are being done to figure out what to display is known as the render step. Each image displayed is a *frame*.

When coding camera behaviors, having the camera update run with the render step produces smoother animations than if you used `for` loops to do the same thing.

You can bind functions to run with each render step by getting `RunService` and using `BindToRenderStep()`:

```
local RunService = game:GetService("RunService")
RunService:BindToRenderStep("Binding Name", 1, functionToBind)
```

In order, the parameters are the following:

▶ **Name:** The name of this specific binding. You need this so you can unbind the function when you want it to stop.

▶ **Priority:** How soon in the render step it happens. The standard is for player inputs to be in the 100s and camera controls in the 200s. If you're unsure, you can the enum `RenderPriority` (see the example script).

▶ **Function:** The name of the function you want to bind.

Offsetting the Camera

If you need the camera to move relative to the player, the Humanoid has a really useful camera property called `CameraOffset`. Examples of when to modify it include creating a bobble effect as the player walks or a camera shake when the player touches something dangerous.

`CameraOffset` takes a `Vector3` and typically needs to be called from the client:

```
humanoid.CameraOffset = Vector3.new(x, y, z)
```

Like other code dealing with camera properties, `CameraOffset` can only be used on the client side. Here, we're showing you how to send a signal from the server to the client to create a camera shake. Camera shakes are classically used when a player touches something harmful, to show the weight of giant monsters, or when the player has been involved in a crash. Here, we're creating a simple hazard part and using the camera shake to give the players feedback that they've touched something dangerous.

1. In ReplicatedStorage, add a RemoteEvent named HazardEvent.

2. Create a new part named Hazard and add your code. The following is basic code for demonstration purposes, but you can use what you learned about module scripts in Hour 19 to set up a more future-proof variation.

```
-- Checks for player touch. If true, subtracts 20 health
local hazard = script.Parent

local function onTouch(otherPart)
    local character = otherPart.Parent
    local humanoid = character:FindFirstChildWhichIsA("Humanoid")

    if humanoid then
        local currentHealth = humanoid.Health
        humanoid.Health = currentHealth - 20
        hazard:Destroy()
    end
end

hazard.Touched:Connect(onTouch)
```

3. Get the necessary services, the RemoteEvent, and then `FireClient()`:

```
-- Checks for player touch. If true, subtracts 20 health
local Players = game:GetService("Players")
local ReplicatedStorage = game:GetService("ReplicatedStorage")

local hazardEvent = ReplicatedStorage:WaitForChild("HazardEvent")
```

```
local hazard = script.Parent

local function onTouch(otherPart)
    local character = otherPart.Parent
    local humanoid = character:FindFirstChildWhichIsA("Humanoid")
    local player = Players:GetPlayerFromCharacter(character)


    if humanoid then
        hazardEvent:FireClient(player)
        local currentHealth = humanoid.Health
        humanoid.Health = currentHealth - 20
        hazard:Destroy()
    end
end


hazard.Touched:Connect(onTouch)
```

4. In StarterPlayerScripts, add a LocalScript and the following variables. Take notice of how the character is checked; you don't want the code to run if the player's character has not fully loaded.

```
-- Services
local ReplicatedStorage = game:GetService("ReplicatedStorage")
local RunService = game:GetService("RunService")
local Players = game:GetService("Players")

local hazardEvent = ReplicatedStorage:WaitForChild("HazardEvent")
local player = Players.LocalPlayer

local character = player.Character
if not character or not character.Parent then -- Check for character
    character = player.CharacterAdded:wait()
end
local humanoid = character:WaitForChild("Humanoid")
local random = Random.new()

local SHAKE_DURATION = 0.3 --How long the shake will last
```

5. Set up the function that generates a random number for the x, y, and z values to be used by CameraOffset.

```
-- Generates random values for CameraOffset
local function onUpdate()
    local x = random:NextNumber(-1, 1)
    local y = random:NextNumber(-1, 1)
    local z = random:NextNumber(-1, 1)
    humanoid.CameraOffset = Vector3.new(x,y,z)
end
```

6. Add a new function to bind `onUpdate()` to the render step for the length of `SHAKE_DURATION`, then unbind it:

```
-- Connect and then disconnect from the render step
local function shakeCamera()
    RunService:BindToRenderStep("CameraShake", Enum.RenderPriority.Camera.
        Value, onUpdate)
    wait(SHAKE_DURATION)
    RunService:UnbindFromRenderStep("CameraShake")
end
hazardEvent.OnClientEvent:Connect(shakeCamera)
```

Connecting to Render Step Indefinitely

If you don't need to control when in the render step the code runs and won't need to disconnect from the render step, the code can be connected to the `RenderStepped` event. The `RenderStepped` event fires *before* the frame is rendered. To avoid affecting game performance, be careful to not connect too many things to the `RenderStepped` event. You can connect it to the same way you can with any other event.

```
local RunService = game:GetService("RunService")
-- Code
RunService.RenderStepped:Connect(functionName)
```

Practice using the tweenservice in combination with the runservice to create the type of camera spin you might see in a loading screen or in a game trailer (Figure 20.7).

FIGURE 20.7
Camera view rotating around the pink sphere to show off the horizon.

Use the following steps:

1. Create a part for the camera to rotate around.

2. In StarterPlayerScripts, add a new local script and the following variables:

```
-- Rotates camera around an object
local RunService = game:GetService("RunService")

local focus = workspace.Focus -- Change to your part
local focalPoint = focus.Position
local camera = workspace.CurrentCamera
camera.CameraType = Enum.CameraType.Scriptable
local angle = 0
```

3. Add the following function to rotate the camera around the part:

```
local function onRenderStep()
    local cameraPosition = focalPoint + Vector3.new(50 * math.cos(angle), 20,
        50 * math.sin(angle))
    camera.CFrame = CFrame.new(cameraPosition, focalPoint)
    angle = angle + math.rad(.25)
end
```

4. Connect the function to the RenderStepped event:

```
RunService.RenderStepped:Connect(onRenderStep)
```

▼ TRY IT YOURSELF

Changing the Time

See if you can change the amount of time it takes for the camera to fully rotate around the part.

deltaTime

Another thing to keep in mind is that the time it takes for every machine to complete the render step is different. Faster devices are capable of refreshing more often than lower-end devices.

Because you don't know exactly how long each frame will take, you can use deltaTime to check that your function is running for the intended amount of time. deltaTime checks the amount of time that has passed between events:

```
local RunService = game:GetService("RunService")

local function checkDelta(deltaTime)
        -- Print the time since the last render step
```

```
print("Time since last render step:", deltaTime)
end

RunService:BindToRenderStep("Check delta", Enum.RenderPriority.First.Value,
    checkDelta)
```

TRY IT YOURSELF ▼

Using `deltaTime`

See if you can make the previous script run for a specific amount of time using `deltaTime`.

TIP

Troubleshooting

If you find that your code isn't working, you may need to change the priority for when it is evaluated in the render step. Another possibility is that not everything the script needs has been loaded in correctly.

Summary

Changing the camera's default behaviors can change how the players experience the game. Camera manipulation can give the players feedback as they interact with objects and create cinematic moments. The camera has properties such as FieldofView that can be modified; however, most modifications are done through code.

Camera code is typically run on the client side using LocalScripts. The camera can be controlled by getting the `CurrentCamera` object, and making it scriptable. Remote events can be used to send signals from the server side to the client side. Always remember to return `CurrentCamera` to custom when the script finishes.

For code that needs to run any duration of time, use the render step rather than a loop. This method is more reliable and typically results in smoother camera motion.

The render step for every device takes a different amount of time. You can check how long the render step takes by using `deltaTime`.

Q&A

Q. Does the camera have any visual 3D representation?

A. No, but it does have a 3D position (Camera.CFrame).

Q. What does the camera render?

A. The 3D world.

Q. What's the default CameraSubject?

A. The Humanoid.

Q. How do you remove built-in behavior from the camera?

A. Set your CameraType to scriptable.

Workshop

Quiz

1. What's the purpose of CameraType?
2. List three camera properties.
3. What is the amount of time during which calculations to refresh the screen are made?
4. What are `BindToRenderStep()`'s three parameters in order?

Answers

1. CameraType controls the behavior of the Camera—that is, how it interacts with the world and subject.
2. CFrame, CameraType, and Focus are three camera properties.
3. The render step is the amount of time during which calculations to refresh the screen are made.
4. The three parameters of BindToRenderStep are Name, Priority, and Function.

Exercises

Use what you know about camera movement to show off your game and create an engaging trailer that can be played on the game page and on social media. Consider these tips:

▶ Scripts can be disabled in Properties (Figure 20.8) if they get in the way of you recording.

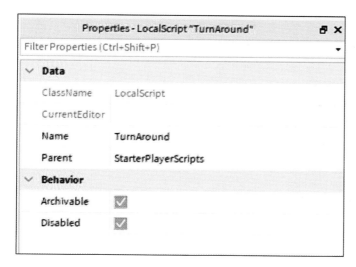

FIGURE 20.8
LocalScript Properties window showing Disabled as checked.

▶ Tweens can be triggered one after another by detecting when the last tween has completed:

```
tween1.Completed:Connect(functionName)
```

▶ Use camera spins to show off the areas of your game that you're most proud of.

▶ Free screen capture software such as OBS can be used even if you don't have more expensive editing suites.

In this second exercise, look for a place within your game that can be improved with a bit of camera movement, for example:

▶ Using a shake while chopping a tree

▶ Pointing the camera at a talking NPC

▶ Add camera motion that shows off your environment the very first time a player loads

HOUR 21
Cross-Platform Building

What You'll Learn in This Hour:

- ▶ What the best practices are for optimizing performance
- ▶ How to make your game mobile-compatible
- ▶ How to test mobile compatibility using Roblox Studio tools
- ▶ What additional steps to take for Xbox/VR compatibility

A game is nothing without its players, and there are some things that you may want to keep in mind to allow the maximum number of people to be able to play your game and for each player to have a more enjoyable experience in-game. In this hour, you learn the best practices to optimize game performance, how to build your game for mobile compatibility, and what things to consider for console and virtual reality (VR) support.

Improving Game Performance

The best part about Roblox is that players can start playing on their computers and then continue playing on their mobile devices with near seamless transition. However, the difference in hardware between devices can drastically hinder the experience if the game is not properly optimized. There are some things that you can do in your game that will ensure that the greatest number of devices will be able to play.

Memory Usage

The main issue for performance is memory use. The developer console tells players how much memory (measured in Megabytes, or MB) is being used, which is necessary data for Roblox developers to optimize their performance.

To access the developer console (Figure 21.1), press the F9 key on your keyboard or simply click the Roblox logo at the top-left corner. Go into Settings and scroll down to open the console.

FIGURE 21.1
Developer console.

Although devices are getting better over time, the benchmark goal for memory usage will be around 700 to 800 MB of memory, to accommodate for lower-end mobile devices. You may use more if you plan to stick to PC and console platforms, but later we examine why mobile is worth consideration, too.

Optimizing Builds

Every part that you place in your game takes a certain amount of memory. It's important to keep in mind *how* each and every part is used to ensure the least amount of memory is being taken up by your game existing on the player's device.

Part Count

Part count is simple: Keep it as low as possible. As you add more parts into the game, Roblox tries to keep up and render each one, as well as the accompanying physics. Beyond limiting how many parts you insert into the workspace, you also can try to optimize parts in the game by either creating a *union* or replacing the part with a *mesh* (read more in the next section). Unions and meshes reduce the amount of physics that Roblox needs to do, and using a union or mesh also reduce the performance strain of part count in your game.

Unions and Meshes

Unions are simpler to create because you can create them within Roblox Studio. Use the following steps to create a union:

1. Highlight all the parts that you want to combine.

2. In the Model tab, under the Solid Modeling category, click Union (Figure 21.2).

FIGURE 21.2
Solid Modeling tools under the Model tab.

3. If you want to undo it in the future, simply click Separate.

Meshes are a little more complicated because creating them requires external 3D modeling software, such as Blender. The process is largely the same:

1. Highlight the parts that you want to combine.

2. Right-click in Explorer and select Export Selection.

3. Save it somewhere on your computer as a .obj file.

4. Open the file in your modeling software of choice.

5. Once you've exported the .obj file, import it back into Studio by adding a MeshPart, clicking the little folder icon next to the MeshId property (Figure 21.3), and opening the file.

FIGURE 21.3
The MeshId property and the icon for importing .obj files.

Why do unions and meshes reduce memory usage? Well, instead of Roblox rendering and calculating the physics for a set of objects, it only needs to do that for one. You can take it one step further by also changing the RenderFidelity property of either unions or meshparts to Automatic. This allows Roblox to cut down on memory for you by not rendering details of parts when the player is far away.

NOTE

Don't Union/Mesh Everything

Unions and meshes, like parts, still use memory, so don't overuse them. There's also a chance that unions or meshes may actually be *worse* for performance if they have a high triangle count (usually more than 5,000). Check with your modeling software or in Roblox Studio (unions have a TriangleCount property) to ensure that unions or meshes are the right fit.

Reusing Meshes and Textures

There are some other things that you can do to optimize mesh performance. Due to the way that the Roblox engine works, meshes that you reuse (and don't change) have a noticeable benefit on performance compared to using different meshes for everything. Therefore, creating one mesh and reusing it many times throughout your game may reduce the memory usage of Roblox. For example, instead of making a unique door for every house, why not just use a couple?

The same goes for textures. Instead of having a dedicated texture file for every single mesh, why not use the same texture, such as the brick walls of a house, shine for windows, and wood grain for the deck, in different, appropriate places?

Reducing Physics

One of the reasons why parts can take up so much memory is because of the physics that needs to be calculated on them every single frame. You can help the physics engine by changing some properties in each part to reduce what needs to be done.

One such property is Anchored. Simply put, a part that is anchored will not move from its position, and so does not need physics to be calculated. Another way to cut down on the physics is changing the CanCollide and CollisionFidelity (if applicable) to a part. CanCollide determines whether a part can, well, collide with other parts. Turning this off means that even if a part does collide with it, nothing will happen. CollisionFidelity changes the collision box of the part to anything ranging from a box around the entire object to one that matches what the object looks like. Using an option such as Box, as shown in Figure 21.4, makes for faster physics calculations than a more detailed CollisionFidelity option.

FIGURE 21.4
The Anchored, CanCollide, and CollisionFidelity properties can all be found on MeshParts and Unions.

Streaming the Content

One of the easiest ways that you can optimize a game is to use Roblox's content-streaming feature. Instead of everything being loaded in and rendered at once, Roblox can choose to show the player only what is closest to them, thereby reducing the memory cost.

However, this option is not for every game. Unloading content means that if a script needs a specific part that isn't loaded in, an error occurs. Streaming can also be suboptimal for games that teleport players because areas may need to be loaded and unloaded quickly. In fact, your players could *fall through the map* if you're not careful! Make sure to check that your game is ready for streaming if you plan to take this option. Once you're ready, enable the StreamingEnabled property in the Workspace (Figure 21.5).

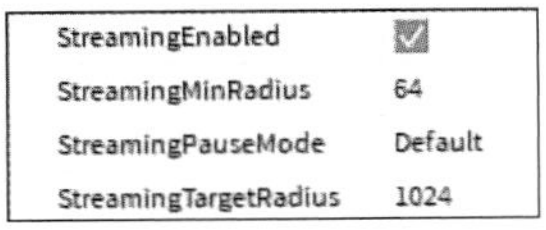

StreamingEnabled	☑
StreamingMinRadius	64
StreamingPauseMode	Default
StreamingTargetRadius	1024

FIGURE 21.5
StreamingEnabled and some properties related to content streaming.

Miscellaneous Tweaks

Beyond the parts themselves, there are some other things that you can try to increase performance. Here they are, in no particular order:

▶ Changing the lighting from ShadowMap to Voxel, which you can find as the Technology property of the Lighting object in the Explorer. Voxel generally creates fewer shadows than ShadowMap, which improves overall performance.

▶ Remove the ability for parts to cast shadows altogether by disabling the CastShadow property. With this disabled, parts no longer cast shadows, no matter what lighting mode your game is set to, which reduces the rendering load.

▶ Delete what players cannot see to reduce lag. Examples include the bottom-side of the terrain or the backside of a building. Solid terrain is fine. Don't try to hollow out things like mountains or volcanos. However, if you have a large terrain block with caves and tunnels, the game slows down because those objects are being accounted for.

▶ Try to avoid using a Transparency value between 0.1 and 0.9 because there are some optimizations that Roblox uses internally that are not possible with an object that is not fully transparent or fully opaque.

▶ In the same vein, minimize using part transparency higher than zero. Although fully transparent parts are not rendered at all, partially transparent objects can have significant rendering costs. Having many translucent parts may slow down the game's performance. However, sometimes you need to use transparency—for example, a large group of windows. In this case, it's better to use one larger sheet of glass rather than individual parts for each window. Your game performance will be improved by making the most use possible out of the fewest possible semitransparent parts.

TRY IT YOURSELF ▼

Optimize Your Game's Parts!

Whether it's combining them into a union, anchoring them, or even flipping the switch to enable StreamingEnabled, see if you can reduce the amount of memory your game uses by optimizing the parts that your game uses! Make sure to keep track of how much it uses before and after.

Improving Your Scripts

Besides the appearance of the game, how it functions can also play a big role in determining how much memory it uses. Although it may seem simple to script and accomplish a task, how much time and how much memory it takes can often make even the most trivial of tasks, such as sorting a list, much more complicated.

Setting the Parent of Objects

Even something as simple as setting the parent of an object too early can slow down your game. When an object is created using `Instance.new()`, the default parent is nil, or nothing (Figure 21.6).

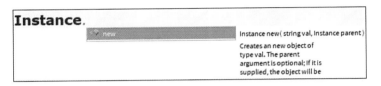

FIGURE 21.6
Instance.new default parent is nil.

If you have been reading the description that Roblox provides as you type something, you might have noticed that the function can take a second argument: the parent of the new object. When an item is parented, Roblox begins to listen for changes to certain properties. Instead of parenting the new part to the workspace using the parameter, parent it only after other changes, such as location and color, are updated. In the following code snippet, parts are created and parented using both methods just discussed. You can see in Figure 21.7 that the time elapsed when running the code for FasterPart is less than the time elapsed for NewPart.

```
local NewPart = Instance.new("Part", game.workspace)
NewPart.Size = Vector3.new(3,3,3)

local FasterPart = Instance.new("Part")
FasterPart.Size = Vector3.new(3,3,3)
```

```
Time elapsed (NewPart): 4.6099999963189e-05
Time elapsed (FasterPart): 2.9300000278454e-05
```

FIGURE 21.7
Showing time elapsed creating each part.

Overreliance on Server/Client

Creating a good multiplayer experience requires a sense of balance of what to put on the server side and what to put on the client side. Rely too much on the client for logic, and your game can become an easy target for exploiters. Rely too much on the server to run everything, and players who have poor Internet connections could have a bad time. Worse yet, overload the server, and you have the potential to lag out *everyone*—not just one person.

Generally, the server is dedicated to running the *logic* of the game server This could be anything from shops to damage and making sure that the information coming in from players is correct.

On the other hand, the client is dedicated to running the *look* and *feel* of a game. This ranges from animations to sounds and even to the user interface. Although this makes it easier for exploiters to manipulate these things, the downside is outweighed by the reduced load on the server.

Use Loops Sparingly

When it comes to game logic, quite a few things could be checked using loops, and whether you're simply going over a group once or every frame, it takes time and memory to do so. Here are a couple of suggestions for what you can do instead of using a loop or, if you need one, how to cut down on the cost of doing so:

- ▶ Using an object (such as a character, part, and so on). Consider the GetPropertyChangedSignal event to run a function when a specific property has changed to avoid constantly checking. Events are your friend in this case, and Roblox has many of them to choose from that listen for pretty much anything you need to account for. Make sure to check the wiki before settling for a loop.

- ▶ Instead of checking *all* items in a list, try to check a specific one.

- ▶ Avoid using loops to fire RemoteEvents or RemoteFunctions. Sending too much information can create a backup of requests, which in turn makes actions in game *feel* slower.

- ▶ Avoid using loops to create large numbers of new objects, *especially parts*. Many changes in a short amount of time, especially when it comes to the workspace, can lead to slowdowns.

Making Your Game Mobile-Friendly

As a Roblox developer, not making your game mobile-friendly can amount to a huge loss in potential players. In fact, according to article featuring Matt Curtis, vice president of developer relations at Roblox, more than *51%* of Roblox players are on mobile, with 44% on PCs, and 5% on consoles (https://venturebeat.com/2020/05/02/roblox-believes-user-generated-content-will-bring-us-the-metaverse/). It's important not only to consider launching on mobile devices but also to ensure that those who play on those devices will have a good experience.

Looks

When it comes to designing for mobile, one of the biggest areas of frustration can be scaling the UI properly. It *is* possible to get away with most modern devices simply by using scale and offset, as they typically have a screen resolution of 1980×1080. However, not all devices will be supported. One of the ways that you can make your UI fit on nearly every screen is by using the UIAspectRatioConstraint (Figure 21.8). This object automatically changes the size of your UI object based on a specified aspect ratio. This ratio can be calculated by simply dividing the *width* of your UI object by its *height*.

FIGURE 21.8
A UIAspectRatioConstraint located within a frame.

Controls

Desktop players have the mouse and keyboard, whereas mobile players are largely restricted to virtual buttons on their screen. That means that a lot of controls that we typically take for granted are not available, at least by default. There are some things that you can do to make it easier for mobile players to play your game.

One such way to remedy this issue is through the usage of ContextActionService. While you may already be familiar with UserInputService, binding an action allows for greater control of what needs to be pressed/tapped, and when it can be done. To get started, simply get the service as you would any other, and use the BindAction function to specify the name of the function, what the action does, whether an onscreen button will be created for touch devices, and the keyboard/controller button input. Figure 21.9 shows a game that has had onscreen features added for this functionality.

FIGURE 21.9
A screenshot of *Arsenal* by ROLVe. Note the added buttons for emotes, reloading, swapping weapons, and shooting the gun.

The following is a code sample of using ContextActionService. Note the `true`, or third argument, determines whether a touch screen button will be created:

```
Local ContextActionService = game:GetService("ContextActionsService")
Local function ActionFunction()
End
ContextActionService:BindAction("Action", ActionFunction, true, Enum.KeyCode.H,
    Enum.KeyCode.ButtonX)
```

Once the button is created, it is placed inside the PlayerGui container, under the ContextAction-Gui screen GUI object (Figure 21.10). From there, you can customize the buttons how you want.

NOTE

Must Use Device Simulator

The features only show up if you are testing with a simulated mobile device, which we go over next.

FIGURE 21.10
The location of the newly created button from ContextActionService.

Simulating Mobile Devices

Fortunately, Roblox Studio allows for easy testing for mobile devices, even if you don't happen to have one on hand. Under the Test tab, in the Emulation category, you can click Device (Figure 21.11) to see your game emulated on a different device.

FIGURE 21.11
Device testing under the Emulation category in the Test tab.

You will also find the device's name, screen resolution, and memory in the middle of the screen just above your workspace. Click it to change the device that is being emulated or even build your own device by selecting Manage Devices to open the Emulation Device Manager shown in Figure 21.12.

FIGURE 21.12
The Emulation Device Manager window.

It's important for you to test your game on many screen resolutions to ensure that it looks good across a wide variety of devices. You can also see how well the game runs with the simulated memory.

NOTE

Switch Tabs to Simulate Devices

If your device emulation option is grayed out, it might be because you're still accessing a script. Switch back to the workspace tab (the tab where you can see all the parts and objects in your game), and it should allow you to continue.

▼ TRY IT YOURSELF

Make Your Game Mobile-Friendly

Use the following tips to make your game mobile-friendly:

▶ Add some buttons to accommodate for touch screens.

▶ Scale the UI to fit every screen.

▶ Test it all with device emulation to ensure that players who play on mobile have the best experience possible.

▶ Make sure to test on a wide range of devices and that it runs well on most of them too.

Finally, try playing the game yourself on a mobile device to make sure that everything works well. Emulation is good, but it doesn't beat testing on a real device.

Console and VR

Although not a big part of Roblox's current player base, designing your game to be console-friendly and potentially VR-compatible will become more and more important as more players try out Roblox on these platforms.

Xbox Guidelines

Although you can instantly publish your game to desktop and mobile platforms, publishing to a console requires you to first acknowledge that your game follows the guidelines for releasing on Xbox. Many of these guidelines are also guidelines of publishing games on Roblox in general—for example, no gambling or encouragement of doing so, no profanity, and so on—but breaking one of these may result in your game not being accessible to players on Xbox. Figure 21.13 shows a list of guidelines for publishing on Xbox.

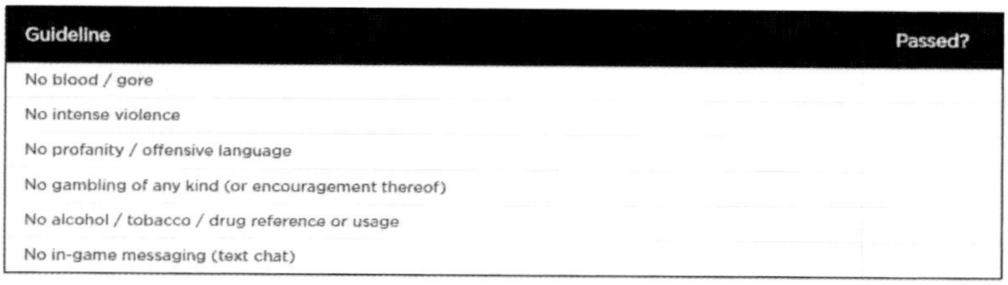

Guideline	Passed?
No blood / gore	
No intense violence	
No profanity / offensive language	
No gambling of any kind (or encouragement thereof)	
No alcohol / tobacco / drug reference or usage	
No in-game messaging (text chat)	

FIGURE 21.13
Guidelines for launching your game on Xbox.

Furthermore, there is another set of guidelines that developers must follow if they want their game to be *featured* on Xbox. This list is much more intensive, and contains many more requirements based on how the game *feels* to play on top of the other requirements. For example, game controls are always responsive unless the game is loading, core game loop only ends when the player wants to or when they reach a fail state, and so on. Due to the sheer number of items to check off, I highly recommend checking out the article on the Developer Hub (https://developer.roblox.com/en-us/articles/xbox-guidelines) if you want to pursue this path of getting featured on Xbox.

VR Best Practices

Virtual reality is still a niche area, but it has been becoming more and more popular across the world, and Roblox is no exception. Developing VR games has its own set of challenges, but if done right, VR can reward the player with a deeply immersive and comfortable experience.

Moving the character is one of the most fundamental parts of playing a game, and if done incorrectly, it can potentially cause the player to become nauseated in real life. Generally, slow and constant motion will be a better experience than sudden jolts or movement in a direction where the player is not looking. There is also an option to *forgo movement altogether* and instead allow the player to teleport from place to place.

Visuals and audio also play a big role in player enjoyment, especially in VR. Placing sounds around the player and playing them as expected (for example, a player may hear a door creak when it's opened) will greatly increase immersion. UI that is drawn in 3D space, such as placing the player's health bar on an in-game watch, can help make the experience more comfortable compared to simply sticking with a flat, 2D UI that is plastered onto the screen. The player may also feel more comfortable if the objects around them are proportional to their size in real life, and so you might need to do some work in scaling objects to the correct size in relation to the character.

▼ TRY IT YOURSELF

Make Your Game Console- or VR-Ready

Although it's not as big of a market as mobile, ensuring that the largest number of players are able to enjoy your game on a variety of devices is an important step toward becoming a great developer. Run down the checklist for Xbox or jump into the game yourself and test it out!

Summary

In this hour, you've learned about optimizing your game for a wide range of devices. You've learned to slim down the memory usage of your game through building and scripting, and you've been introduced to some potential next steps on future projects, such as designing around StreamingEnabled. You've also learned about the mobile player base and how to accommodate for them through the use of onscreen buttons and proper UI scaling. Finally, you've learned about guidelines and best practices for bringing your experience onto Xbox or taking immersion to the next level through virtual reality.

Q&A

Q. Is it possible to change what devices my game supports after releasing it?

A. Yes! Under the Home tab, within the Settings category, you can configure your game settings. From there, under Basic Info, you can mix and match what devices your game supports at any time.

Workshop

Now that you have finished, take a few moments to review to see if you can answer the following questions.

Quiz

1. True or False: The main goal for developers is to reduce *memory usage*.

2. To reduce the load that Roblox has on parts, you can combine groups of them into ____ and ____.

3. One easy way to improve performance (that can be risky) is to enable the _____ property and stream the content instead.

4. True or False: It is better to set the *parent* of the object first before setting its properties.

5. True or False: More than 50% of Roblox's player base plays on mobile devices.

6. One way to create buttons for mobile devices is through the _____.

7. True or False: There are few guidelines to getting your game featured on Xbox.

8. One way to properly scale the UI for mobile devices is by adding a _____.

Answers

1. True. Memory usage plays a big factor in what devices can play our games.

2. To reduce the load that Roblox has on parts, you can combine groups of them into unions and meshes.

3. One easy way to improve performance (that can be risky) is to enable the StreamingEnabled property and stream the content instead.

4. False. Setting the *parent* of an object before setting its properties can actually cause performance issues.

5. True. About 51% of players play on mobile devices.

6. One way to create buttons for mobile devices is through the ContextActionService.

7. False. There are many guidelines to follow before your game can be featured on Xbox.

8. One way to properly scale the UI for mobile devices is by adding a UIAspectRatioConstraint.

Exercises

This exercise will combine a number of different things you've learned this hour. If you get stuck, don't forget to refer to the previous pages in this hour. Make a button that does something and is compatible with both desktop and mobile.

1. Create a LocalScript inside *StarterPlayerScripts*.

2. Inside the script, create a function that does whatever you want.

3. Get the ContextActionService and bind the function to the action. Make sure to set the third argument to `true` for the service to create a button for you.

4. Test the game by emulating a mobile device of your choice. Make sure to try pressing the button to see if it works.

5. **Extra:** Customize the button by changing its properties in the script. Check the section for mobile UI if you're unsure where the button is located in the Explorer.

Bonus Exercise: Publish the ultimate game by optimizing it, making it mobile-ready, and releasing it to every device (including console).

1. Check your parts. Are they all properly anchored? Would it be possible to reduce the memory usage by replacing some with meshes or unions? Check the Developer Console, Memory, Place Memory to make sure that no one area is taking up *too* much memory.

2. Check your scripts. Are they all optimized? Would it be possible to get the same results/do the same thing without using a loop?

3. Test your game using the device emulator—is it possible to do everything that players on PCs can do on mobile? Is the UI scaled properly?

4. Test your game using a real mobile device. Are the controls intuitive, or does anything need tweaking to improve the experience?

5. Go over the Xbox release guidelines on the Roblox Developer Wiki, and make sure that your game passes all of them.

6. Go into your game settings and release your game on all possible platforms.

7. **Extra:** Go over the Xbox featured guidelines and make sure that your game passes those too. Good luck!

HOUR 22
Global Community Building

What You'll Learn in This Hour:

▶ What localization is
▶ How to adapt your game for international audiences
▶ How to modify your game for the global compliance system
▶ How to adhere to data privacy laws

Roblox players come from countries all around the world and have different backgrounds, cultures, and expectations. In this hour, you find out how to adapt your game for different players to make sure that your game is following the global compliance system, and adhering to the increasing number of data privacy laws that are being created across the globe.

Introduction to Localization

Localization, at its core, is making sure that all players who play your game feel welcome. This can range from something as simple as translating the game to their native language to something more complex, such as mixing up entire sections of the game to better fit within a certain culture. Although this is an extra step that may not contribute to your core game, it is not only important for the players but also to your game because localization can allow more players (who may not speak English or understand American culture) to enjoy the experience that you have created.

Capturing Text for Translation

Translation tools are offered to developers who want to manually translate their games and upload them or allow others to help translate their game. Let's first go over how to translate all the text from your game. Use the following steps to begin working on translating your game:

1. Go to your game on the Roblox website, and click the ellipsis button in the upper right (Figure 22.1).

FIGURE 22.1
Home page of *Beat the Scammers!* by Roblox Resources.

2. Click Configure Localization (Figure 22.2).

FIGURE 22.2
The drop-down menu for games, including the Configure Localization button.

3. Go to the Settings tab on the left and enable Automatic Text Capture (Figure 22.3) to allow Roblox to collect text automatically when it shows up in-game whether it's on your screen on in the world itself.

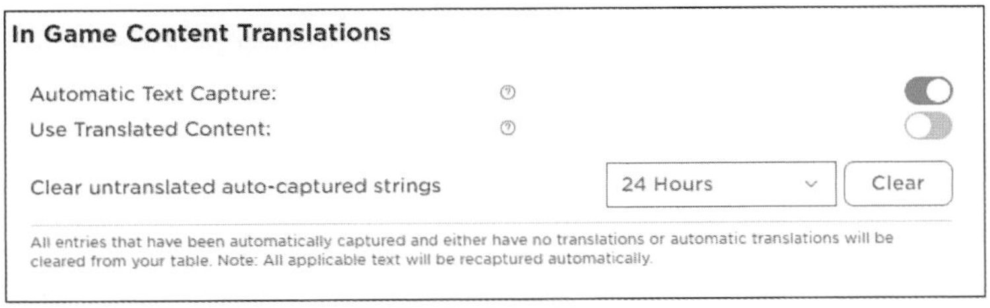

In Game Content Translations		
Automatic Text Capture:		
Use Translated Content:		
Clear untranslated auto-captured strings	24 Hours	Clear

All entries that have been automatically captured and either have no translations or automatic translations will be cleared from your table. Note: All applicable text will be recaptured automatically.

FIGURE 22.3
The In Game Content Translations screen.

4. Go back and play the live version (not the Studio file) of your game for a minute or two.

5. To see the text that has been captured, go back to the Configure Localization page from step 1 and add a new language by typing in the name of a supported language in the Add Language box (Figure 22.4).

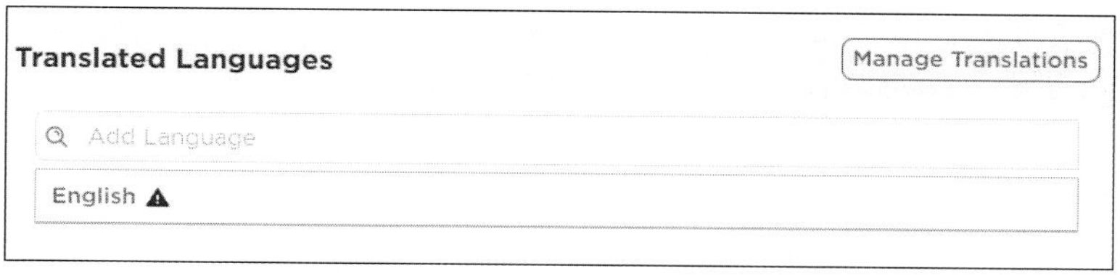

Translated Languages   [Manage Translations]

🔍 Add Language

English ⚠

FIGURE 22.4
The main screen after clicking Configure Localizations.

Translating Captured Text

Now that you have all of the text in your game captured, let's take a look at how to translate them on the website. You start by clicking the new language that you added in the previous section, which takes you to the Manage Translations page (Figure 22.5).

FIGURE 22.5
The main Manage Translations page.

There are a couple of main sections covered here:

▶ The list of source text strings is underneath the button marked Add New Entry. It contains a list of all the strings that have been captured.

▶ The box on the right labeled Enter Translation Here is the text box where you will enter the translation for the text.

▶ There are several tabs underneath the translation box that tell you where the text is in-game (Locations in Game), as well as previous translations (Translation History).

Deploying Your Translations

Once your text is translated and ready to be seen by your international players, go back to the Settings tab in the Configure Localization page and enable the Use Translated Content option (Figure 22.6). Once that is enabled, you will be able to see the translated content while playtesting in Roblox Studio as well as in-game with the correct settings.

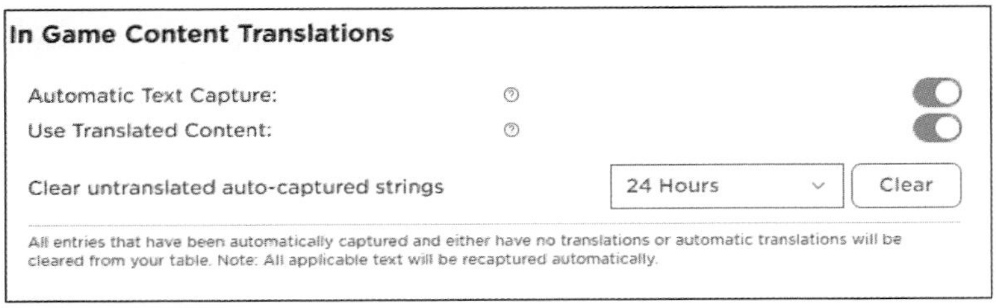

In Game Content Translations

Automatic Text Capture:	⑦	
Use Translated Content:	⑦	

| Clear untranslated auto-captured strings | 24 Hours ⌄ | Clear |

All entries that have been automatically captured and either have no translations or automatic translations will be cleared from your table. Note: All applicable text will be recaptured automatically.

FIGURE 22.6
Enabling Use Translated Content so that international players can see translated text.

Although translating seems simple enough, there are a couple things to keep in mind:

▶ Translated text may be longer than your source, so make sure to allow for extra space.

▶ Avoid using parameters, such as currency count or lives, because they can get messy when translated. These are shown with curly braces ({}) inside the Manage Translations tab.

▶ Be careful when using icons/UI elements with text as an image, such as an ImageButton for a shop where the word "Shop" is part of the image. You'll need to manually replace those images of UI elements because Roblox does not capture or translate them.

Hire a Translator

While it may seem tempting to use existing translation software such as Google Translate to translate your game to a different language, it is often better to hire someone who speaks the language that you need. Automated translating has been announced on Roblox, but is not yet available to the general public. Even if it were generally available, a lot of things can be lost with automation. Hiring someone to translate your game can provide a better experience and replace some cultural references with ones that local people may better understand.

Global Compliance

As Roblox expands overseas, different cultures and local laws have made it necessary to introduce a new global compliance system for developers to follow, especially if they are interested in releasing their games to certain regions. As such, Roblox has introduced a function called GetPolicyInfoForPlayerAsync (Figure 22.7) that returns a list of policies that a player should follow.

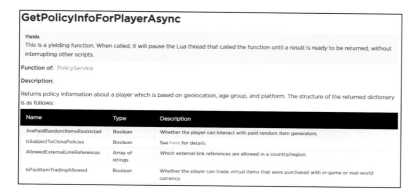

GetPolicyInfoForPlayerAsync

Yields
This is a yielding function. When called, it will pause the Lua thread that called the function until a result is ready to be returned, without interrupting other scripts.

Function of: PolicyService

Description:

Returns policy information about a player which is based on geolocation, age group, and platform. The structure of the returned dictionary is as follows:

Name	Type	Description
ArePaidRandomItemsRestricted	Boolean	Whether the player can interact with paid random item generators.
IsSubjectToChinaPolicies	Boolean	See here for details.
AllowedExternalLinkReferences	Array of strings	Which external link references are allowed in a country/region.
IsPaidItemTradingAllowed	Boolean	Whether the player can trade virtual items that were purchased with in-game or real-world currency.

FIGURE 22.7
The GetPolicyInfoForPlayerAsync function returns a dictionary of policies a player may be restricted to.

Although it is ultimately up to the developer to implement these policies, failure to do so may result in a game being temporarily or permanently removed from the platform. These policies are

▶ **ArePaidRandomItemsRestricted:** Often known as "loot boxes," this policy determines whether a player from a certain area can use real currency (Robux) to open them.

▶ **IsSubjectToChinaPolicies:** This one is a bit more abstract. Not much information about this is available, but we can assume that games on Roblox must follow other games that want to be released in China. This includes translating *all* content to Simplified Chinese, explicitly showing chances of obtaining an item in a random item generator (loot box), and limited or no gore/blood.

- ▶ **AllowedExternalLinkReferences:** This policy restricts what websites a player can visit outside of Roblox. For example, a player may not be able to visit a developer's Twitter profile if Twitter is prohibited in their country.

- ▶ **IsPaidItemTradingAllowed:** This policy determines whether a player will be allowed to trade items that were either obtained with in-game money or real-world currency (Robux).

Although there is no one main way to account for policies, an easy set of steps and guidelines to follow would be if a game element has

- ▶ Loot boxes/paid random items
- ▶ A link to an outside website
- ▶ Item trading
- ▶ Gore/blood
- ▶ Gambling

Privacy Policies: GDPR, CCPA, and You

As more countries around the world begin to adopt laws that give the consumer/player more control over their data, it is time for developers to also adapt and ensure they could honor this request. Developers may need to establish a system in their game to accommodate this need. Under the General Data Protection Regulation (GDPR), as well as the California Consumer Privacy Act (CCPA), both Roblox and you are barred from storing certain types of information and must delete that information should the person in question file a formal request.

General Guidelines

While these two laws affect different regions (EU and California respectively), they share similar characteristics in what types of information cannot be stored. As a developer, you must avoid collecting *personal information* about your players, which includes things such as birth dates, personal photos, email addresses, and more. If you have done so already, make sure to change your system so as not to store that information in the future.

If a request for data deletion is made, make sure that it is done through Roblox official channels (they may send you an email ending in the roblox.com domain). If a player asks for their data to be removed from your game, ask them to first send a data deletion request to Roblox Support at https://www.roblox.com/support.

Removing Player Data

In the situation that you may need to remove player data from your game, it's nice to have a system in place already. In our example, all the data that a player has is stored in one key labeled "Player_" followed by their user ID. As such, removing that key would also remove all their data from the game at once. Follow these steps:

1. Begin by placing a BindableEvent inside ServerStorage (Figure 22.8). This serves as your way of signaling Roblox to remove a given player's data.

FIGURE 22.8
A BindableEvent inside ServerStorage.

2. Place a script inside ServerScriptService (Figure 22.9). This holds the function that deletes a given player's data when given the signal to do so.

FIGURE 22.9
A script inside ServerScriptService.

3. Insert the contents of the following script. Once the signal has been received, the function attempts to get the player's data from the DataStoreService in the format of Player_UserID, where "UserID" is the actual user ID of the player. If found, the data is removed. Otherwise, the error message prints what went wrong. This can range from Roblox data stores being down to the player data simply not existing.

```
local ServerStorage = game:GetService("ServerStorage")
local DataStoreService = game:GetService("DataStoreService")
local removePlayerDataEvent = ServerStorage:WaitForChild("RemovePlayerData")

-- Reference to player data store (replace "PlayerData" with the name of your
    data store)
local playerData = DataStoreService:GetDataStore("PlayerData")

local function onRemovePlayerDataEvent(userId)
        -- Pattern for data store player key, for instance "Player_12345678"
        local dataStoreKey = "Player_" .. userID

        local success, err = pcall(function()
            return playerData:RemoveAsync(dataStoreKey)
        end)
```

```
        if success then
            warn("Removed player data for user ID '" .. userID .. "'")
        else
            warn(err)
        end
    end

removePlayerDataEvent.Event:Connect(onRemovePlayerDataEvent)
```

To run this script and delete the player's data:

1. From the View tab in Studio, open the Command Bar (Figure 22.10).

FIGURE 22.10
Use the Command Bar to execute commands to the server.

2. With everything set up, you can run Studio with zero players by clicking the little blue arrow underneath Play in the Home tab (Figure 22.11).

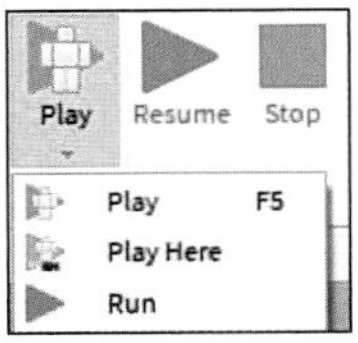

FIGURE 22.11
Use Play to run Studio with no players.

The game launches on the server side.

3. When you have the game running and the console out (you should see a text box at the bottom of your window), you can type in the command in Figure 22.12 and run it by pressing Enter. If your data stores have data for that player, it should be removed.

```
game:GetService("ServerStorage").RemovePlayerData:Fire("Player_505306092")
```

FIGURE 22.12
We will be running this command in the console.

Summary

In this hour, you've learned about expanding your community to encompass international players and localizing your game through translation. We also shared some guidelines that you may want to follow when releasing games to certain countries. Finally, you've learned about new privacy guidelines and how to set up your game in case a data deletion request is filed.

Q&A

Q. Why do I need to follow guidelines for data in other countries?

A. The short answer is that to do business in EU countries or in California, companies and developers must follow the local regulations. Although you may live outside of both areas, you have to follow their guidelines if you want players from that area to play your game.

Workshop

Now that you have finished this hour, take a few moments to review and see if you can answer the following questions.

Quiz

1. True or False: Translation can be done locally or through the cloud.

2. The function that allows developers to check what policies a player has is called _____.

3. The alternative (and more commonly known) name for random item generators is _____.

4. True or False: To have a game published in certain countries, you must comply with their regulations.

5. True or False: Living outside the EU or California allows you to avoid implementing data protection policies.

6. You can execute functions in Roblox Studio through the _____.

7. True or False: Manual translation is generally better than automated translation.

8. You can remove stored player information by erasing it from the _____.

Answers

1. True. You can modify the spreadsheet manually or use the Roblox website translation portal to translate games.

2. The function that allows developers to check what policies a player has is called GetPolicyInfoForPlayerAsync.

3. The alternative (and more commonly known) name for random item generators is loot boxes.

4. True. For countries with stricter regulations, you must follow guidelines to release your game there.

5. False. If you want to conduct business in the EU or California, you must implement the GDPR/CCPA.

6. You can execute functions in Roblox Studio through the Command Bar.

7. True. Grammar, cultural references, and slang can often go unnoticed by automated translation software.

8. You can remove stored player information by erasing it from the data store.

Exercises

This exercise combines a number of different things you've learned this hour. If you get stuck, don't forget to refer to the previous pages. In this exercise, you check to make sure that your game is ready for international release through a combination of translations, policy checks, and systems!

1. Has your game been automatically or manually translated to another language?

2. Does your game have checks for players who may be restricted from purchasing loot boxes with real money or trading?

3. Is your game free from anything that may collect the *personal information* of players?

4. Does your game have a system in place that can delete player data at Roblox's request?

5. **Extra:** Ask around your community to see if any international players have feedback! They will offer the best information on what to do to localize for their regions.

HOUR 23
Monetization

What You'll Learn in This Hour:

▶ How to implement one-time purchases using Game Passes

▶ What Developer Products are and how to use them to make consumables

▶ What Roblox Premium is and how to monetize it

▶ What the Developer Exchange program is and how to use it

So, you've come up with a great game idea, or maybe you have a game that's nearly finished, and you want to earn some Robux from it? In this hour, you'll learn how you can monetize your game to earn Robux, and in turn, use the Developer Exchange program to convert your hard-earned Robux into real money! You'll learn how to develop one-time purchases using Game Passes, how to increase revenue with consumables, and what Roblox premium is and how to use it in monetization. Finally, this hour wraps up with a look at the Developer Exchange program.

Game Passes: One-Time Purchases

Game Passes on Roblox are for items that a player can purchase only one time in a game, such as for perks like a VIP vehicle or double XP. Items the player acquires with a game pass belong to that player permanently.

One-time purchases offer a couple of benefits that make them stand out from other forms of monetization. First, one-time purchases, especially at a low cost, can introduce the player to spending in your game. Once a player has bought one product and is invested in the game, so to speak, they are more likely to go on and buy more products. (And if they start buying consumables and engaging in repeat purchasing, that's even better!) Second, once they are invested in the game, they are also likely to play more, which increases retention and average playtime, hopefully pushing your game closer to the front page!

It's time to try making a one-time purchase product! Start off by creating a Game Pass on the Roblox website. Click the Create tab on the website, and then find your game by clicking the My Creations or Group Creations tab. Then click Games (Figure 23.1).

FIGURE 23.1
Click the Games tab.

On the right side of the screen, click the Settings button and choose Create Game Pass (Figure 23.2).

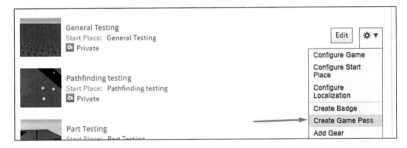

FIGURE 23.2
Click the Create Game Pass button.

You have the option to upload a picture for your Game Pass (Figure 23.3). Photos need to be a 150 × 150 circle; you can find online templates for sizing images. You'll also need to give the Game Pass a name (such as "Double XP" or "VIP") and a description. Once you're done filling in the information, click the Preview button (Figure 23.3).

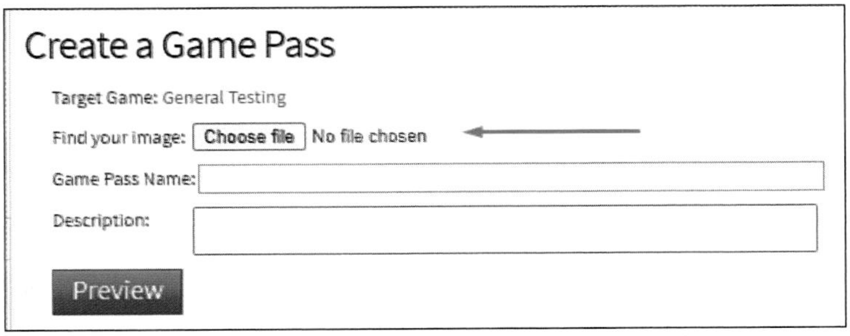

Create a Game Pass

Target Game: General Testing

Find your image: [Choose file] No file chosen

Game Pass Name: [   ]

Description: [   ]

[Preview]

FIGURE 23.3
Upload an image.

You can then preview the Game Pass to make sure you're happy with how it looks before clicking Verify Upload.

Once you've created the Game Pass, you should see it in the Game Passes tab. Click the Settings button for it and then click Configure. The Configure Game Pass dialog box (see Figure 23.4) opens, and you can choose a price. Give some thought to the purchase price for Robux and consider how much you think the item is worth in comparison to other products you are selling. It can often be helpful to compare your game to other games of a similar genre and check out their products so you know how much other developers are selling those items for. After you've entered the price for your Game Pass, click the Save button.

FIGURE 23.4
Setting a price for your Game Pass.

Selling Your Game Pass in Game

So, now you have a Game Pass on sale, which a player can buy from the website (on the Game Passes tab). It would be much more convenient for the player, however, if you prompted purchase of the product in your game (and you would likely sell a lot more Game Passes). Thankfully, Roblox has a way to do that! By using `MarketPlaceService`, you can prompt a Game Pass purchase using `PromptGamePassPurchase`, like so:

```
:PromptGamePassPurchase(player, ID)
```

For example, you can prompt the purchase when the player clicks on a `TextButton`. The following code is used to prompt a Game Pass purchase:

```
local MarketplaceService = game:GetService("MarketplaceService")
local ID = 12345678 - Get the ID of the Game Pass from the URL

MarketplaceService:PromptGamePassPurchase(game.Players.LocalPlayer, ID)
-- Prompt Game Pass purchase to the LocalPlayer (client)
```

You need to get the ID of the Game Pass to use in this function call. Click the Game Pass on the website to get the ID from the URL of the page, as shown in Figure 23.5.

FIGURE 23.5
The Game Pass ID in the URL.

Enabling Game Pass Perks

When you first set up Game Pass in-game purchasing, after a player buys the Game Pass, nothing changes in the game until you give the Game Pass a perk. Do this by using `UserOwnsGame-PassAsync` to see whether the player owns the Game Pass. If they do, you can then give them the perk that was originally advertised using code like this:

```
local userID = game.Players.LocalPlayer.UserId
local ID = 1234567 -- ID of the Game Pass

if MarketplaceService:UserOwnsGamePassAsync(userID, ID) then
    -- Player owns this Game Pass
end
```

When the player joins the game, you can check for a Game Pass. However, you don't want to check continuously to determine whether the player buys the Game Pass while in the game. To solve this, you can use the `PromptGamePassPurchaseFinished` event, which fires when the player either closes the prompt or buys the Game Pass. If the player purchases the Game Pass while in the game, you can then award the perks just like before, like so:

```
function FinishedGamePassPrompt(player, purchaseID, purchased)
    if purchased == true and purchaseID == ID then
        -- Player bought Game Pass with the ID checked
    end
end

-- This will call the "FinishedGamePassPrompt" function once the player either buys
    the Game Pass or closes the prompt
MarketplaceService.PromptGamePassPurchaseFinished:Connect(FinishedGamePassPrompt)
```

NOTE

Pending Sales

Every time something is purchased from your game, the revenue from that product sits in Pending Sales (Figure 23.6) for three days before it's released into your account (or group). This perfectly normal precaution is designed to prevent scams!

FIGURE 23.6
The Pending Sales indicator.

Developer Products: Consumables

Game Passes are a great way to monetize your game. However, repeat purchases (also known as consumables) can really increase your revenue by encouraging players to spend money multiple times in your game! This is where Developer Products are really useful. Developer Products—or "dev products" as they're more commonly known—are used to implement consumables in the form of repeat purchases. For example, they can be used to sell limited-time powerups or extra in-game currency. You should never create a dev product that is too similar to a Game Pass. For example, if you create a dev product that gives 2x walkspeed for 15 minutes and a Game Pass that gives permanent 2x walkspeed, no one will buy the dev product!

To create a Developer Product, go to your game's page, click the three-dot button, and then click Configure This Game (Figure 23.7).

FIGURE 23.7
Click Configure this Game.

On the Configure Game page, click Create New (Figure 23.8) and fill in the price and name for your product. You'll also be able to upload an icon for the dev product.

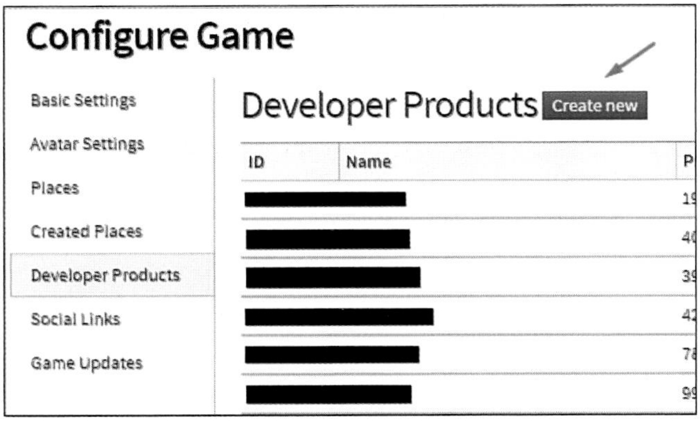

FIGURE 23.8
Create a new Developer Product.

Just like the Game Pass, you can provide the player with an in-game prompt to purchase a dev product. To do this, use `PromptProductPurchase`. To create the prompt, you need to copy the ID of the Developer Product from the website, as shown in the following code:

```
local productid = 1234567

local player = game.Players.LocalPlayer
MarketplaceService:PromptProductPurchase(player, productid)
```

Now, this is where it gets slightly tricky. But don't worry; we're breaking it down to make it easy to understand. Whenever a player purchases a Developer Product, the transaction needs to be handled using something called the `process receipt` callback. This callback will be called whenever a player purchases a Developer Product in your game:

```
local MarketplaceService = game:GetService("MarketplaceService")

function processReceipt(receiptInfo)
    -- Handle purchasing
end

-- The processReceipt function will be called everytime a purchase is made in the
    game
MarketplaceService.ProcessReceipt = processReceipt
```

Inside the `processReceipt` function, you need to award the player the item/perk they purchased. First, however, check that the player is still in the game. If they have left, you should return `Enum.ProductPurchaseDecision.NotProcessedYet` so Roblox can automatically refund the Robux to the player, or you can award the perk the next time the player joins. If you choose the latter, the `processReceipt` callback will be called again the next time the player joins, or, if they don't join in the next three days, the Robux will be automatically refunded to them.

The most important part is returning `Enum.ProductPurchaseDecision.PurchaseGranted`, as shown in the following code. If you do not return this at the end of the `processReceipt` function, you won't receive the Robux from the sale!

```
local MarketplaceService = game:GetService("MarketplaceService")

function processReceipt(receiptInfo)
    local player = game:GetService("Players"):GetPlayerByUserId(receiptInfo.
        PlayerId)
    if not player then
        -- The player left the game
        -- If they come back, the callback will be called again
        return Enum.ProductPurchaseDecision.NotProcessedYet
    else
        -- Award the product and if all goes well, PurchaseGranted can be called
    end

    -- This MUST be called otherwise you won't receive Rob
    -- Very important it is only called when the purchase is successful, however
    return Enum.ProductPurchaseDecision.PurchaseGranted
end

-- The processReceipt function will be called every time a purchase is made in the
    game
MarketplaceService.ProcessReceipt = processReceipt
```

NOTE

Marketplace Fee

Whenever you sell a dev product or Game Pass, Roblox will take 30% of the revenue earned. This is known as the "marketplace fee."

Roblox Premium

If you're already a keen Roblox player or developer, you're probably aware of Roblox Premium, a subscription service that gives players extra Robux every month and enables them to trade items on the Roblox catalog. A unique way to monetize your game is to give players perks for being a Premium member, which in turn encourages them to buy Roblox Premium. You may be thinking, "Surely that just benefits Roblox?" However, Roblox recently introduced something called Premium Payouts (Figure 23.9), which enable you to earn Robux based on how many Premium players play your game and how long they play. So, by encouraging players to purchase Premium by giving perks in your game, you can increase the Premium Payouts you receive.

NOTE

Roblox Premium

Roblox Premium is also needed for DevEx, which we talk about in the next section of this hour.

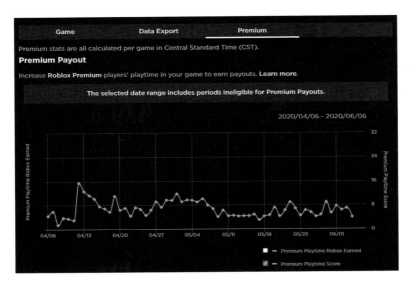

FIGURE 23.9
Premium Payout graph on the developer stats page.

To implement perks, you can check a player's Premium membership status using the following code:

```
local player = game.Players.LocalPlayer
if player.MembershipType == Enum.MembershipType.Premium then
     -- This player owns Roblox Premium!
end
```

To prompt players to purchase Premium, you can use PromptPremiumPurchase:

```
local player = game.Players.LocalPlayer
if player.MembershipType == Enum.MembershipType.Premium then
     -- This player owns Roblox Premium!
else
     MarketplaceService:PromptPremiumPurchase(player)
end
```

Similarly, with Game Passes, to handle situations where a player purchases Premium while playing your game, you should use the `PlayerMembershipChanged` event to enable the relevant rewards:

```
local Players = game:GetService("Players")

-- Event will be fired when a player's membership changes
Players.PlayerMembershipChanged:Connect(function(player)
    if player.MembershipType == Enum.MembershipType.Premium then
        -- Player just bought Premium!
    end
end)
```

The server will be waiting for a reply from the client. If the client is laggy, it may be waiting a long time. If the client disconnects/leaves the game, the function will error. If you do use an `InvokeClient`, you should wrap the function in a pcall!

TIP

Avoid Making Your Game "Pay to Win"!

"Pay to Win" is where a product in your game gives an unfair advantage to paying players. The model will put off players who don't have any Robux because they will think your game is unfair and therefore not fun for them.

Developer Exchange: Earn Real Money from Your Game

Developer Exchange (DevEx) enables developers to convert Robux into real money. Since it was introduced in 2013, developers have been paid millions of dollars and have been able to form companies using the revenue generated from their games. In theory, DevEx is available to everyone, but there are some requirements you should be aware of:

- ▶ You have to be a member of Roblox Premium.

- ▶ You have to have a minimum of 100,000 Robux in your account.

- ▶ You must have a verified email address.

- ▶ You must be 13 years of age or older.

- ▶ You must be a good community member who abides by the terms of service (TOS). (Roblox will check your moderation history to see if you have been previously warned or banned. If you have a clean moderation history and are a good member of the Roblox community, you should have no problems!)

Assuming you meet all these requirements, DevEx allows you to submit a request each month for real money. The current rate at the time of writing is 100,000 Robux for $350 USD—this amount is converted into the appropriate currency for your country. You are able to receive the payment through PayPal, as a bank transfer, or, in some cases, as a check.

To cash out with DevEx, go to the Create tab and select Developer Exchange. Click the big Cash Out button on the right (Figure 23.10) and fill in all the relevant details. If it's your first time DevExing, you should receive an invite to the Tipalti portal, which is how Roblox pays you. If not, it can take up to two weeks to receive the money; however, in most cases, payments are made much quicker than that most of the time.

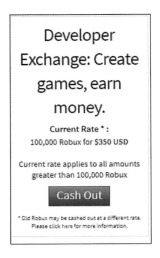

FIGURE 23.10
The Cash Out button.

There is no limit to how many Robux you can cash out (Figure 23.11) at one time.

FIGURE 23.11
On the DevEx page, you can fill out how many Robux you want to cash out.

NOTE

What You Can Cash Out—And What You Can't

The Robux you cash out must have been earned from your games or clothing—it can't be from selling Limiteds on the Roblox catalog.

Summary

In this hour, you learned the basics of monetizing your game on Roblox. You've been introduced to how to make and implement Game Passes to create one-time purchases in your game and how to use Developer Products to create consumables, which can be purchased multiple times. We also covered Roblox Premium and how you can use it to increase revenue. And finally, we talked about the Developer Exchange program and how you use it to convert Robux into real cash!

Q&A

Q. What is Tipalti, which was mentioned in the DevEx section?

A. Tipalti is a payment processor that Roblox uses to pay people. You can log in to the online portal to view payment details and payment history, as well as submit tax forms and more.

Q. Should I use Game Passes or dev products to monetize my game?

A. You should use both! Having a mixture of one-time purchases and repeat purchases encourages different types of players to spend while also offering the opportunity to develop a continuous revenue stream from repeat purchases. This can also help convert players to whales (the big spenders in a game), which helps you generate more revenue.

NOTE

Offer Less to Get More

When implementing monetization into your game, you should be careful about adding *too* many products. If you have a lot of products to choose from, players will be overwhelmed. Think about the type of products that players will *want* to buy, and price them appropriately; try to have a variety of prices to target different players and types of spenders.

Workshop

Now that you have finished this hour, take a few moments to review to see if you can answer the following questions.

Quiz

1. What's the minimum amount of Robux you can cash out using DevEx?

2. ____ are the players who spend the most in your game.

3. Game Passes are an example of ____-time purchases.

4. True or False: You can only buy dev products once.

5. True or False: You get revenue from Roblox Premium purchases in your game.

6. True or False: If a player purchases a dev product in your game, you automatically get the Robux.

7. True or False: Revenue from your game takes three days to arrive in your account or group.

Answers

1. 100,000 is the minimum amount of Robux you can cash out using DevEx.

2. Whales are the players who spend the most in your game.

3. Game Passes are an example of one-time purchases.

4. False. They are used for repeat purchases.

5. False. You do get Premium Payouts for engagement time of Premium players, however.

6. False. You *must* return a `PurchaseDecision` in the `ProcessReceipt` function to receive funds.

7. True. Robux earnings sit in Pending Sales for three days before being released into your account or group.

Exercises

This exercise combines a number of different things you've learned this hour. If you get stuck, don't forget to refer to the previous pages in this chapter. Create a Game Pass that gives the player extra walkspeed. This is a permanent perk because players own Game Passes forever!

1. Create a simple UI with an `ImageLabel` showing the Game Pass image; a `TextLabel` describing the perk the player will get, along with the price; and a `TextButton` for prompting the player to make a purchase.

2. In the `TextButton`, add a `LocalScript` that prompts the player to purchase the Game Pass. Don't forget to use the correct ID from the Game Pass URL.

3. In `ServerScriptService`, create a script that uses the `PromptGamePassPurchaseFinished` event to verify whether the player bought the Game Pass. If they did, increase the walkspeed of the player (Step 5).

4. Add a `PlayerAdded` event to the same `Script`, which checks, every time a player joins, whether they own the Game Pass. If they do, increase the walkspeed of the player—this can be done by calling a function so both the `PromptGamePassPurchaseFinished` and `PlayerAdded` event can use it.

5. Create a function called `increaseWalkspeed`, which takes a player parameter. Use `Player.Character.Humanoid.Walkspeed = 50` to increase the walkspeed. You can set the speed to any value you'd like.

Bonus Exercise: Create a Developer Product that increases the walkspeed of the player every time they purchase it. (For the sake of simplicity, we won't use any data stores to save this value, but ideally you would need to save purchases/perks so they can be awarded the next time the player joins.)

1. Create a simple UI with an `ImageLabel` showing the dev product icon for the product, a `TextLabel` describing that the dev product will award the player, and a `TextButton` to prompt the purchase.

2. In the `TextButton`, insert a `LocalScript` that prompts the player to purchase the dev product. You'll need to use the ID from the dev product you created on the website.

3. In `ServerScriptService`, create a script that uses the `Marketplaceservice.ProcessReceipt` event to verify which dev product the player bought (check the ID) and to make sure they're still in the game.

4. If the player is still in the game and the ID matches, increase the walkspeed and return the `PurchaseGranted` decision.

5. If the player has left the game, you should return `NotProcessedYet`. If the player is still in the game but the ID matches, they have bought another dev product in your game, and you should handle that appropriately.

6. You can reuse the same `increaseWalkspeed` function you made in the first exercise. This time, however, you should increase the walkspeed by a certain amount each time. For example, `Player.Character.Humanoid.Walkspeed = Player.Character.Humanoid.Walkspeed + 10`.

7. **Extra:** Using data stores, try and save a purchase history log for each player!

HOUR 24
Attracting Players

What You'll Learn in This Hour:

▶ How to make the most of game thumbnails, icons, and trailers
▶ How to keep a game fresh with updates
▶ How to put advertising and notifications to use
▶ How to understand player activity with analytics

Roblox gives you several ways to lure players to your game and to help make your game successful. This hour discusses how you can use game icons, thumbnails, and trailers to give players an exciting first glimpse at your game. It also explains how you can update your game to keep it interesting and fresh to returning and new players, and how you can advertise it to potential new players. The last topic covers how you can use Roblox Analytics to better fine-tune and market your game.

Game Icons, Thumbnails, and Trailers

Game icons and thumbnails work as covers of the game. You must make them as attractive as possible, just as books often have appealing, interesting covers. The icon is especially important because it's the first thing players see, so it should seem enticing to make people want to play your game. After a player clicks your game's icon, its details page should include thumbnail images to further draw players into your game. Figure 24.1 shows games' icons on the Roblox home page, and Figure 24.2 shows a thumbnail on the details page of a game.

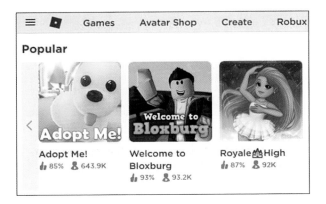

FIGURE 24.1
Game icons on the Roblox home page. (Games shown: *Adopt Me!* By DreamCraft, *Welcome to Bloxburg* by Coeptus, and *Royale High* by callmehbob.)

FIGURE 24.2
Game thumbnail on the details page for *Royale High* by callmehbob.

In addition to game thumbnails and icons, you also have the option to add a trailer video of your game to show off the game's features to prospective players.

Thumbnails, icons, and video trailers represent your game to the world. One game can have up to 10 thumbnails, 1 icon, and 1 video trailer. Most video trailers are animated, which allows players to really see all the advantages to your game.

To add a thumbnail, icon, or trailer to your game, do the following:

1. Open your game page or go to the Create tab.

2. On the Create tab, click the gear icon next to the game and then select Configure Start Place from the drop-down menu (Figure 24.3).

FIGURE 24.3
On the Create tab, selecting the Configure Start Place option.

3. Alternatively, if you're on your game page, click the ellipsis button at the top and then select Configure This Place from the drop-down menu (Figure 24.4).

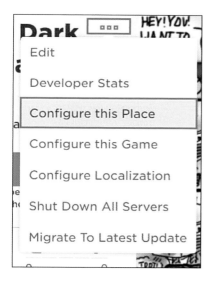

FIGURE 24.4
On the game page, selecting the Configure Start Place.

Regardless of which method you use to select the Configure This Place option, the Configure Place page launches (Figure 24.5).

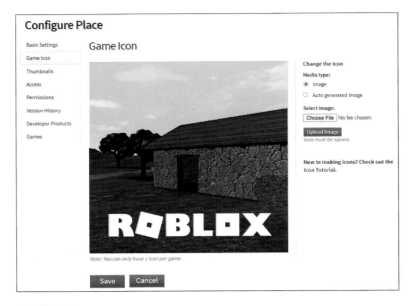

FIGURE 24.5
The Configure Place page.

After the page opens, follow these steps:

1. Click the Game Icon tab on the left.

2. Click the Choose File button. Then pick an image and click the Upload Image button.

3. Click the Save button to add an icon on the game.

If you don't have a specific image from the game that you want to use, you can instead select the Auto Generated Image option, which snaps an image from the game and creates an icon.

To create thumbnail images for the game, use the following steps:

1. Click the Thumbnails tab on the left.

2. Click the Choose File button, find the image you want to use, and click the Upload Image button (Figure 24.6).

3. You can repeat these steps to upload up to 10 thumbnail images for free.

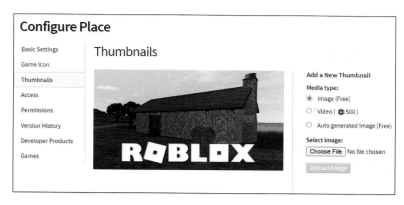

FIGURE 24.6
Uploading thumbnails.

Adding a video trailer for your game works much the same way as adding a thumbnail. On the right of the Thumbnails tab, you select the Video option under Media Type. You have to meet the following requirements to upload a trailer:

▶ You must pay 500 Robux to upload a video trailer, but doing so can enormously help with the marketing of your game.

▶ The trailer must be 30 seconds or shorter. If it's longer than 30 seconds, you receive an error.

▶ You must upload the trailer to YouTube.

After you select Video as the Media Type on the Thumbnails tab, copy a YouTube link (URL) into the appropriate field and click the Add Video button. Click the Save button when you're done. The Roblox Moderation Team reviews the video, and then it will be posted to the game page.

Updates

To make games successful, attractive, and continually functional, you must keep updating your game. If you never update the game, players may lose interest in it, but regular updates—whether that's every week, month, or season—can make the game seem fresh. For example, if your game involves car racing, examples of updates include updating the UI (user interface), adding new maps or updating the current ones, replacing objects, or improving game performance.

NOTE

Seasonal Updates

Here are some ideas to update your games for each season: In winter, you can add snowfall and snow-covered trees to your game. In summer, you can change the game theme to hot and sunny and change the Skybox to sunny weather, and so on. If you have terrain on the ground, you can simply go to Terrain, pick leaves or snow, and start painting on the ground. However, if you're promoting seasonal updates, remember that seasons are different throughout the world.

Advertising and Notifications

Advertising your game is also important. If you don't advertise your game, there's a chance that your game won't be successful. Advertising must be really good to attract people, and then the thumbnails will attract the people who came from the advertisement.

There are two main kinds of advertising: sponsor and user ads. For both, you have to bid Robux, but user ads require an ad image. With your ads, you can promote catalog items, games, groups, and models.

Sponsor Ads

Sponsor ads work without ad images, but they do require a bid. (Read more about bidding later in this hour.) They get your game on the Game page, which is also known as Roblox's *front page*, and have a "Sponsored" tag, which can help with getting your game noticed and popular. Here is how you can sponsor your game:

1. Go to the Create page.

2. Click the gear icon and then select Sponsor (Figure 24.7).

FIGURE 24.7
Setting up a sponsor ad.

3. A Sponsor Game page opens. Select which device you would like to sponsor your game on (Figure 24.8). You can select all devices, but keep in mind that you have to separately bid on each of them.

4. When you're done, click Create.

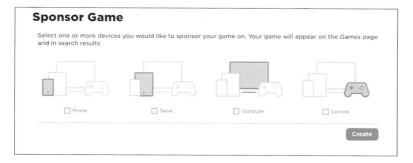

FIGURE 24.8
The menu with different platform choices on which to advertise.

5. You will be directed to Sponsored Games (Figure 24.9), which you also can access yourself by going to the Create tab of Roblox and then clicking the Sponsored Games tab on the left side of the Create page.

FIGURE 24.9
List of Sponsored Games.

6. Click the gear icon next to a game and select Run from the drop-down menu (Figure 24.10) to open a bidding field.

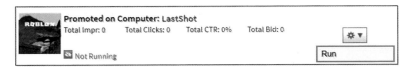

FIGURE 24.10
Clicking the Run option.

7. Type your bidding amount in the Bid in Robux field, and click the Run button (Figure 24.11).

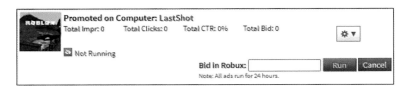

FIGURE 24.11
Bidding for sponsoring games.

NOTE

Bidding Explained

The more Robux you use to bid on the ads, the greater the result. You will receive more clicks, and the ad will display more on Roblox websites. Here's some key terminology about bidding:

▶ An *impression* is every time a player sees your ad. One view is counted as one impression.

▶ A *click* occurs every time a player clicks your ad.

▶ *Click Through Rate* (CTR) is a percentage of players who see your ad and also click it. If all the people who see your ad click it, you have a CTR of 100%.

If you spend 100 Robux on one ad and 300 Robux on another, the 300 Robux bid gets you more impressions and it appears three times as often as the ad for which you bid 100 Robux.

Both sponsored and user ads run for 24 hours. After 24 hours, you have to bid more and click Run to keep the ad running.

User Ads

User ads require an image, and you can use them to advertise catalog items, games, groups, and models. There are three different sizes of ads you can pick:

- **Banner** is sized 728×90 and appears on the top or bottom of the Roblox website.
- **Skyscraper** is sized 160×600 and appears on the left or right side of the Roblox website.
- **Rectangle** is sized 300×250 and appears on the right side or bottom of the Roblox website.

You can download templates for the ad styles from the Create page. Click the gear icon and then select Advertise from the drop-down menu (Figure 24.12).

Edit
Configure Game
Configure Start Place
Configure Localization
Create Badge
Create Game Pass
Add Gear
Developer Stats
Advertise
Sponsor
Shut Down All Servers

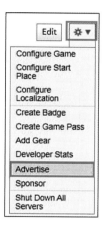

FIGURE 24.12
Access templates for user ads.

The Create User Ad page opens (Figure 24.13), listing all the sizes available. Click your preferred template to download it.

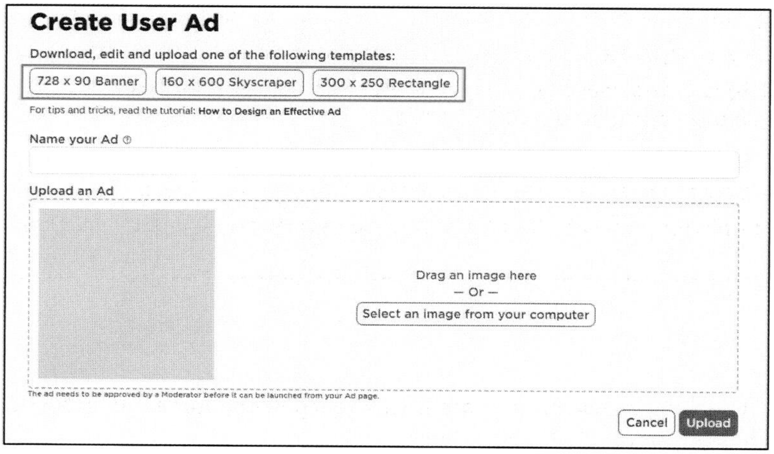

Create User Ad
Download, edit and upload one of the following templates:
728 x 90 Banner
160 x 600 Skyscraper
300 x 250 Rectangle
For tips and tricks, read the tutorial: How to Design an Effective Ad
Name your Ad
Upload an Ad
Drag an image here
— Or —
Select an image from your computer
The ad needs to be approved by a Moderator before it can be launched from your Ad page.
Cancel
Upload

FIGURE 24.13
The Create User Ad webpage.

Once you have your preferred size template downloaded and ready, you need to create your user ad:

1. Go to any model, game, or group you want to advertise.

2. Click the ellipsis or gear icon if it's a game.

3. Click Advertise to open the Create User Ad page (Figure 24.14).

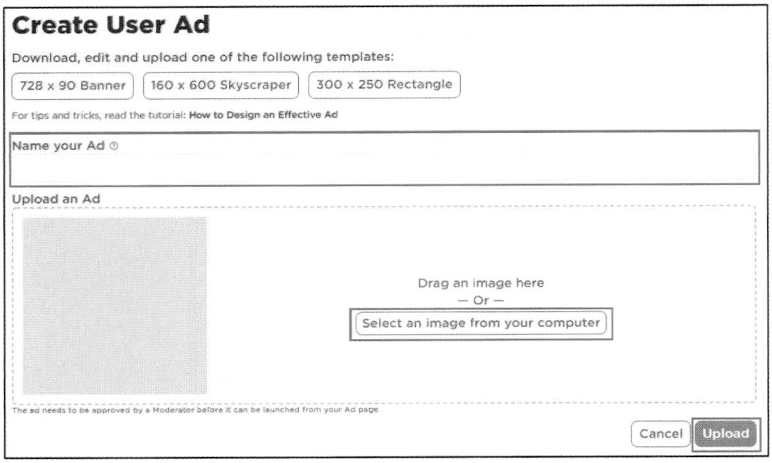

Create User Ad

Download, edit and upload one of the following templates:

[728 x 90 Banner]   [160 x 600 Skyscraper]   [300 x 250 Rectangle]

For tips and tricks, read the tutorial: **How to Design an Effective Ad**

Name your Ad ⓘ

Upload an Ad

Drag an image here
— Or —
[Select an image from your computer]

The ad needs to be approved by a Moderator before it can be launched from your Ad page.

Cancel   **Upload**

FIGURE 24.14
Creating your user ad.

4. Type a name for the ad.

5. Click Select an Image from Your Computer and find the image you want to use.

6. Click the Upload button.

After you've uploaded your ad, the Roblox Moderator will approve it before it appears on your ad page. Once your ad is in your ad page, you can run the ad by doing the following:

1. Open the Create page and click the User Ads tab.

2. Click the gear icon on the right and select Run Ad.

3. Enter a Robux bid amount.

4. Click Run to start the cycle. Every ad runs for 24 hours only.

Notifications

You use notifications to inform the game's followers that you have published an update to the game. You manually send notifications to help drive followers to visit the game to check out the update. You can send only one notification in a week.

Use these steps to send a notification:

1. Go to the Create page and click the gear icon.

2. Select Configure Game from the drop-down menu (Figure 24.15).

FIGURE 24.15
The Configure Game option.

3. Click Game Updates on the left side of the page that opens (Figure 24.16).

FIGURE 24.16
Configuring the game to send notifications.

4. Type a description of the update and then click the Send button.

5. The notification is sent to all the followers of the game.

Notifications appear on the Notifications slot on the Roblox home page toolbar.

Analytics

Analytics keep track of how many players visit the game, what devices they use, Robux Revenue, Premium Payouts, and how many products are being sold in your game. Following is some of the information Roblox Analytics can provide to you:

▶ Number of people visiting your game

▶ Number of hours they are spending playing

▶ Platforms they are using to play your game

▶ Robux Revenue

▶ Premium Payouts

▶ Number of products being sold in your game

This data can be invaluable in updating or marketing your game. For example, using Analytics, you can see how many people were pulled into your game after an update. To open Developer Stats and access Roblox Analytics, do the following:

1. From the Roblox website, go to the Create tab.

2. From the game, click the gear icon and select Developer Stats (Figure 24.17). You see the Developer Stats page where you view all the data points above. Daily, monthly, and yearly data is available to view.

FIGURE 24.17
Accessing Roblox Analytics.

Summary

In this hour, you've learned about game thumbnails, icons, and trailers and how they work to attract people to your game. You have learned about the ways updates can keep the game interesting and generate revenue. You learned about using advertising and notifications to promote your game and send followers notifications about any update you have published, both of which can improve traffic to your game. Finally, you have learned about analytics, which allows you to track players' time in your game, among other data points, to improve your marketing.

Q&A

Q. Is adding thumbnails to the game free?

A. Yes, it's totally free to add thumbnails to a game, but when running ads, you have to bid to run the ad.

Q. How much does it cost to add a video trailer to your game?

A. It costs 500 Robux to add a video trailer to your game.

Q. Will bidding more Robux on ads give you more clicks?

A. Yes, bidding more Robux will give you better results and most likely more clicks.

Q. Does an ad run for 24 hours only?

A. Yes, once you have bid the Robux, it runs for 24 hours only.

Workshop

Now that you have finished, take a few moments to review and see if you can answer the following questions.

Quiz

1. True or False: Do user ads get people to check your game?

2. True or False: Adding an icon to a game is free.

3. True or False: A video trailer has to be uploaded on YouTube to be added to your game.

4. You can add __ thumbnails to your game.

5. True or False: User ads are free to run.

Answers

1. True. User ads do get people to check your game.

2. True. Adding icons to a game is free.

3. True. A video trailer has to be uploaded on YouTube to be added to your game.

4. You can add 10 thumbnails to your game.

5. False. User ads are not free to run.

Exercises

As the owner of a game, planning your release schedule is one of the most important things you can do. It allows you to prioritize your work. For example, a holiday release can often take months of planning, and if you don't plan in advance, you can easily find yourself taken by surprise with not enough time to create a timely update.

Plan a one-year schedule for updating your game. Consider how often you want to update and what will be updated. Include a balance of potential gameplay improvements, additional content, and holiday updates.

1. For each update, write how long you think it will take to achieve.

2. Mark which updates you think are the most important for keeping your players happy and coming back.

3. For each of your most important updates, think about how long they should take to complete and work backward to figure out when you need to begin work on that update to release it on time.

4. For each proposed update, create a list of what you need to create. Include notifications, ads, thumbnails, and videos to keep your players informed.

5. Throughout the year, keep your schedule up to date. Even the best-laid plans need to adjust as surprises pop up. This way, though, you can stay focused on the most important updates while not getting as distracted by smaller ones.

In this second exercise, add a thumbnail to your game.

1. Click the ellipsis button of the game and select Configure Place.

2. On the left side, click Thumbnails.

3. Navigate to your file and click Upload Image; then click the Save button.

The thumbnails must be really good to attract the players to play your game. Figures 24.18 and 24.19 show some examples of the good thumbnails.

FIGURE 24.18
Thumbnails from *Arsenal* by ROLVe Community.

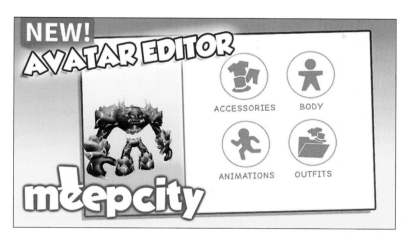

FIGURE 24.19
Thumbnail from *MeepCity* by Alexnewtron showing updated features.

APPENDIX A
Lua Scripting References

In Hour 11, we introduced you to the programming language Lua and provided a brief overview of its foundational concepts. This appendix includes a few additional Lua reference tables and concepts that you may find useful as you begin to study the language.

Modifying Properties That Are Data Type and Enumerations

Data types are the different types of data a variable can store. They are listed in Tables A.1 and A.2.

TABLE A.1 Primitive Lua Data Types

Data Type	Description
Nil	No data.
Boolean	Data is either true or false.
number	Data is real numbers.
string	Data is an array of characters.
function	Data is a method written in C* or Lua.
userdata	C* data.
thread	Data is independent threads of execution.
table	Arrays, symbol tables, sets, records, graph, trees, and so on.

TABLE A.2   Roblox Lua Data Types

Data Type Categories	Custom Roblox Data Types
Colors	BrickColor, Color3, ColorSequence, ColorSequenceKeypoint
Position or Area Related	Axes, CFrame, UDim, UDim2, Rect, Region3, Region3int16
Number and Sequence	NumberRange, NumberSequence, NumberSequenceKeypoint
Connections and Events	RBXScriptConnection, RBXScriptSignal
Vectors	Vector2, Vector2int16, Vector3, Vector3int16
Classes	Instance
Enumeration Related	Enum, EnumItem, Enums
Other Types	DockWidgetPluginGuiInfo, Faces, PathwayPoint, PhysicalProperties, Random, Ray, TweenInfo

Enumerations, or enums, are special data types that store (userdata), a set of values specific to that enum. These are read-only values. To access enums in scripts, you need to use a global object called Enum. You can find the list of enums at https://developer.roblox.com/en-us/api-reference/enum.

Here are some examples of how to change other properties that are a data type or enum now. Let's say you want to create a new part called redBrick with a material of brick and a brick color of red. The part is medium gray stone by default, so to change the color to red, you need to do the following:

```
redBrick.BrickColor = BrickColor.Red()
```

Now to change the material of redBrick to brick, get the materials list from enum because Material is an enum:

```
redBrick.Material = Enum.Material.Brick
```

Notice as you keep typing the code, the editor autosuggests or autocompletes the code for you.

Conditional Structures

Conditional structures are a way to specify the flow control in programs. If a condition is met, Lua treats it as true; if not, the value is either false or nil. These conditionals can be checked by using the relational operators in Table A.3. For the examples, var 1 is 30 and var 2 is 10.

TABLE A.3   Relational Operators

Operator	Description
+	Addition, adds two operands; var1+var2 returns 40.
-	Subtraction, subtracts second operand from first; var1–var2 returns 20.
*	Multiplication, multiplies both operands; var1*var2 returns 300.
/	Division, divides numerator by denominator; var1/var2 returns 3.
%	Modulus, gives the remainder after the division; var1%var2 returns 0.
^	Exponent, will give the exponent value; var1^2 returns 900.

Table A.4 shows the conditional operators. For the examples, var1 is 30 and var2 is 10.

TABLE A.4   Conditional Operators

Operator	Description
==	Equal to, (var1 == var2) is not true.
>	Greater than, (var1 > var2) is true.
<	Lesser than, (var1 < var2) is false.
>=	Greater than or equal to, (var1 >= var2) is true.
<=	Lesser than or equal to, (var1 <= var2) is false.
~=	Not equal to, (var1 ~= var2) is true.

Table A.5 shows the logical operators. For the examples, var1 holds the true logic and var2 holds the false logic.

TABLE A.5   Logical Operators

Operator	Description
and	Logical and, if both operands are non-zero then true; (var1 and var2) is false.
or	Logical or, if any two operands are non-zero then true; (var1 or var2) is true.
not	Logical not, reverses the logical state; !(var1) is false.

Expanding Lua Knowledge

If you want to dive deeper into the Lua programming language, the following two excellent resources are readily available on the Web:

▶ Lua Reference Manual at http://www.lua.org/manual

▶ Programming in Lua at http://www.lua.org/pil

APPENDIX B
Properties and Functions of Humanoid

In Hour 12, the Humanoid object is discussed. The Humanoid includes useful functions and properties that can be used when applying damage, changing display names, manipulating the camera offset, or reading the Humanoid's state (for example, climbing, dead, current health).

Table B.1 shows some of the Humanoid properties. We haven't included all the properties because some are either self-explanatory or not relevant.

TABLE B.1 Humanoid Properties and Meanings

Property	Meaning
Camera Offset	Offsets the camera from the character.
DisplayDistanceType	Viewer: You see everyone's DisplayName at your specified distance.
	Subject: Everyone controls their own specified distance on your PC.
	None: Will not be shown on your PC.
DisplayName	Option to display a custom DisplayName above your character; by default, your username.
HealthDisplayDistance	At what distance you can see others' health bars; default: 100 (Integer).
HealthDisplayType	Options: WhenDamaged, On, and Off.
NameDisplayDistance	At what distance you can see others' names; default: 100 (Integer).
NameOcclusion	OccludeAll: Hide all name tags behind objects.
	NoOcclusion: Name tags are displayed through objects.
	EnemyOcclusion: Hide other team names behind objects.
RigType	R15: 15 body part rig.
	R6: 6 body part rig.
JumpPower	If UseJumpPower: Amount of force applied to jump.
	If not UseJumpPower: Choose height directly.
Jump/PlatformStand/Sit	Booleans, true or false. Can be overridden by player input.

Property	Meaning
TargetPoint/WalkToPart/ WalkToPoint	Primarily used with NPCs and players during cutscenes MoveTo() in scripts: WalkToPoint [First Parameter]. WalkToPart [Second Parameter]. TargetPoint and WalkToPoint use Vector3 inputs. WalkToPart uses Instance input.

Feel free to test out the properties yourself in-game to get a sense of them.

The Humanoid also comes with an array of functions (some of which are shown in Table B.2) for handling things such as equipping tools, loading animations, taking damage, forcing the player to work to X, and setting different player states. We haven't included all the functions because some are either self-explanatory or not relevant.

TABLE B.2   Humanoid Functions and Meanings

Function	Meaning
GetState()/ChangeState()	Fetch or set the HumanoidStateType.
EquipTool()/UnequipTools()	Equips the specified tool.
LoadAnimation()	Loads the specified animation onto the Humanoid that returns an AnimationTrack that can be used to play an animation.
TakeDamage()	Applies damage to the player by lowering their Humanoid.Health property by the given amount.
MoveTo()	Player attempts to walk to a given Vector3 or part; modifies: Humanoid.WalkToPoint/Humanoid.WalkToPart.

Last but not least, the Humanoid has a set of RBXScriptSignals (or events) that you can use for detecting when a Humanoid property changes, among other things. Table B.3 includes a list of the most important events.

TABLE B.3   Humanoid Events and Meanings

Event	Meaning
AnimationPlayed()	Fires when an animation starts playing on the player.
Died()	Fires when the player dies (their health reaches 0 or the Head and Torso become detached).
HealthChanged()	Fires when the player's health is altered.
MoveToFinished()	Fires when the player has been directed to move to a certain position, usually via MoveTo() function.
Seated()	Fires when the player's .Sit property is active.
StateChange()	Fires when the HumanoidStateType changes; parameters: Old & New State.

Relying on events instead of using loops with checks for property changes is good form, and it allows you easy access to firing functions depending on the status of a property and so on while being more performant across the board.

Index